Solve Nonlinear Systems of PDEs by Order Completion :

Can There Be a General Nonlinear PDE Theory for Existence and Regularity of Solutions ?

Elemér E Rosinger

Dedicated to Marie-Louise Nykamp

Table of Contents

Part I : The Main Ideas and Facts	5
Abstract	6
Some of the Most Basic Ideas ... Made Simple ...	7
Brief Sketch of the Book	12
0. A Glimpse into History Related to Science and Paradigms, and the Inescapable ... Competence Rigidity ...	18
1. A Sample of Customary Perception	23
2. The Very Large Class of Nonlinear Systems of PDEs Solved	25
3. A Short History of Difficulties in Solving Linear and Nonlinear PDEs	27
4. Nonlinear Algebraic Theory of Generalized Functions	31
5. The Order Completion Method	32
6. Comparison with Methods in Functional Analysis	34
7. Solving General Equations by Extending their Domains of Definition : the Three Classical Methods	37
8. The Need for Extensions in the Case of Solving PDEs	43
9. Order Completion Abolishes the Dichotomy "Linear Versus Nonlinear"	48
10. The Hidden Power of Methods Based on Partial Orders	50
11. Conclusions	52

12. Appendix : Trying to Ease ... the Pains of
    ... Paradigm Rigidity ... :-) :-) :-)   53

References   65

Part II : Six Papers on the Order Completion Method 73

1. A Brief Announcement about the Increased Blanket
   Regularity of Solutions, and the Definition of
   Their Property   74

2. More Detailed Presentation of the Stronger Blanket
   Regularity Property of Solutions
87
3. Certain Details on Solving Large Classes
   of Nonlinear Systems of PDEs   106

4. A Few General Results on Partial Orders   134

5. Solving Arbitrary Equations by Order Completion :
   Necessary and Sufficient Conditions for the
   Existence of Solutions   148

6. Further Details on Solving PDEs by
   Order Completion   169

Part III : A Few Practical Suggestions ...   192

What to Read First, and How to Try to Read It ...   193

Part I : The Main Ideas and Facts

Abstract

Contrary to widespread perception, there is ever since 1994 a unified, general, that is, type independent theory for the existence and regularity of solutions for very large classes of nonlinear systems of PDEs, with possibly associated initial and/or boundary value problems, see [21,22], and for further developments [1-3,47-56,58,64-66]. This solution method is based on the Dedekind order completion of suitable spaces of piece-wise smooth functions on the Euclidean domains of definition of the respective PDEs. All the solutions obtained have a blanket, minimal regularity property, as they can be assimilated with usual measurable functions or even with Hausdorff continuous functions on the respective Euclidean domains.

It is important to note that the use of the order completion method does *not* require any monotonicity condition on the nonlinear systems of PDEs involved.

One of the major advantages of the order completion method is that it *eliminates* the algebra based dichotomy "linear versus nonlinear" PDEs, treating both cases with equal ease. Furthermore, the order completion method does *not* introduce the dichotomy "monotonous versus non-monotonous" PDEs.

None of the known functional analytic methods can exhibit such a performance, since in addition to topology, such methods are significantly based on algebra, and vector spaces do inevitably differentiate between linear and nonlinear entities.

Some of the Most Basic Ideas ... Made Simple ...

A most general *equation*

(I) $\quad T(x) = c$

means in fact a corresponding *mapping*

(II) $\quad T : X \longrightarrow Y$

where X and Y are certain nonvoid sets, and $c \in Y$ is given. Thus the *problem* is to find a *solution* $x \in X$.
Clearly, the *necessary and sufficient* condition for such a solution $x \in X$ to exist for all $c \in Y$, is that the mapping T be *surjective*.

Equally clearly, if the mapping T is *not* surjective, then for $c \in Y \setminus T(X)$, one can *nevertheless* solve the above equation, if one *suitably extends* the domain X of the mapping T to a *larger* domain $\tilde{X}$, that is, for which we have $X \subsetneq \tilde{X}$. And then, one may try to look for solutions $\tilde{x} \in \tilde{X} \setminus X$.
Here of course, the accent is on the concept of *suitable* !
Anyhow, the mentioned solutions $\tilde{x}$ will then be called *generalized solutions*.

It follows that the mapping T in (II), must be suitably *extended* to a mapping $\tilde{T}$ given by the *commutative diagram*

(III)
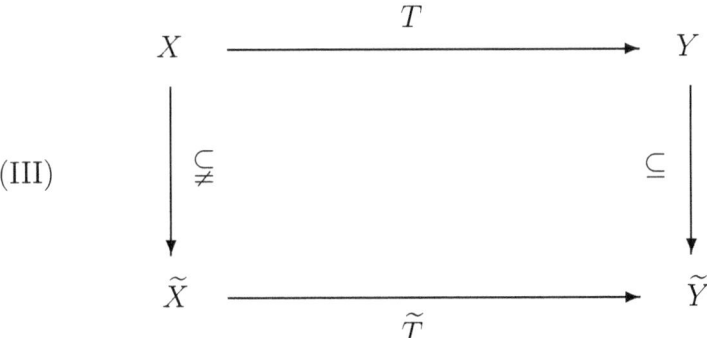

and in general, we may expect that even the set Y may have to be extended to a set $\widetilde{Y}$.

So much for hopefully ... clarifying ... generalities ...

Here however, we shall be interested in solving very general nonlinear systems of PDEs, with possibly associated initial and/or boundary value problems. And it turns out to be useful to start in this regard by considering the above most general related situation. The mentioned very general nonlinear systems of PDEs will have equations such as :

$$\text{(IV)} \quad F(x, U(x), \ldots, D_x^p U(x), \ldots) = f(x), \quad x \in \Omega \subseteq \mathbb{R}^n, \ |p| \leq m$$

where F is any function *jointly continuous* in all its arguments, the right hand term f is also a *continuous* function, the order $m \in \mathbb{N}$ is given arbitrary, while the domain Ω can be any bounded or unbounded open set in $\mathbb{R}^n$. Later, in fact, both F and f will be allowed to have certain discontinuities as well.

Now such PDEs define obviously the following associated mappings :

$$\text{(V)} \quad T : \mathcal{C}^m(\Omega) \ni U \longrightarrow T(U) \in \mathcal{C}^0(\Omega)$$

where

$$\text{(VI)} \quad T(U)(x) = F(x, U(x), \ldots, D_x^p U(x), \ldots), \quad x \in \Omega$$

And now, here comes one of the **basic problems** in solving PDEs like the above ones : in most cases of interest, the corresponding mappings T are **not** *surjective* !

Thus according to the general situation considered at the beginning, we have to ... venture ... into ... *generalized solutions* ...

And this *lack* of surjectivity happens already in the most simple nontrivial cases, like for instance, with linear constant coefficient PDEs. Indeed, the classical *Green function* solutions of such PDEs are definitely generalized functions, in fact, Schwartz distributions. However, they are well known to play a fundamental role even in the study of classical solutions of the respective linear PDEs.

Therefore, from the above most simple and general considerations, it follows that, when solving very general nonlinear PDEs as those in (IV), we may - similar to (III) - have to consider *suitable extensions* of the mappings (V) and (VI), given if possible by *commutative diagrams*

(VII)
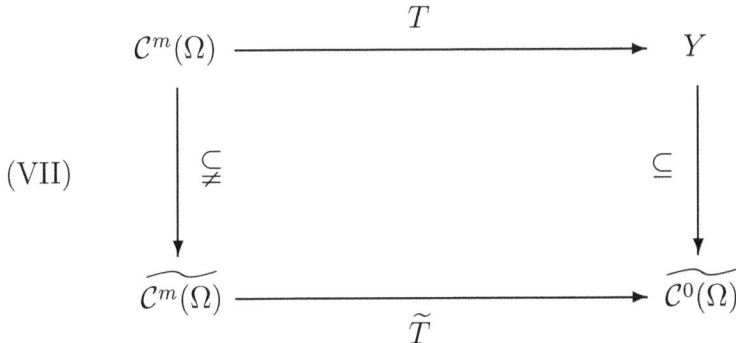

Now, quite fortunately, nothing in the above is new, except perhaps for the somewhat ... simple and general formulation ...

Indeed, regarding the solution of various equations, the stages (I), (II) and (III) have countlessly many times been accomplished, starting more than two and a half millennia ago, when Pythagoras had to find out to his own immense horror, that the equation $x^2 = 2$ did not have any rational number solution, as we formulate his trouble in our modern mathematical terms ...

As for solving, in particular, various linear and nonlinear PDEs, situations of the kind of (VII) have also been encountered many many times, and they do in fact form the basis of countless modern functional analytic methods in solving PDEs.

And then, **what** may the ... *novelty* ... be in this book, that is, with the *order completion* method, when solving very general equations, and in particular, very general nonlinear systems of PDEs, such as those that have equations of the form (IV) ?

Well, the *novelty* with the *order completion* method is, simply, as follows :

The way the *extensions* $X \subsetneq \widetilde{X}$, $\mathcal{C}^m(\Omega) \subsetneq \widetilde{\mathcal{C}^m(\Omega)}$, and so on, are constructed is by the use of *topological completion* of suitable uniform topologies, respectively on X, $\mathcal{C}^m(\Omega)$, and so on.
This is in fact the *essence* of the functional analytic method.

And then, in the above terms, the **novelty**, and thus, the *essence of the order completion* method is that the mentioned *suitable extensions*, such as in (III) and (VII), are obtained this time *not* by use of any topologies at all, but instead, by the use of suitable *natural partial order structures* on the respective spaces of classical smooth functions !

But then, are there certain **advantages** in using the order completion method for solving very general nonlinear systems of PDEs ?

Yes, as seen in the sequel, there **are** indeed, several rather unprecedented such advantages.
Here, briefly we mention a few of them :

- Due to the more *basic* nature of the partial order concept, than of the concepts of algebra or topology, the order completion method simply does *not* differentiate between linear and nonlinear PDEs, solving both cases in a unified manner.

- The order completion method allows the solution of initial and/or boundary value problems associated with PDEs to be solved as well. This is in sharp contrast to the ways functional analytic methods deal with such problems, ways which bring in considerable additional difficulties due , among others, to the fact that the operation of restricting generalized functions to lower level manifolds than the domain of definition of the respective PDEs is well known to be a highly singular one.

- As seen in paper 5 in Part II, surprisingly general equations - far beyond PDEs - can be solved by the order completion method. Furthermore, one can obtain *necessary and sufficient* conditions for their solvability, as well as explicit expressions for their solutions, when they exist.

Further details about the surprising advantages of the order completion method, when solving PDEs, are mentioned in the rest of Part I, and can also be noted in the six selected papers in Part II.

Needless to say, more details can be found as well in [21,22,1-3,47-56,58,64-66].

However and at least so far, for a more thorough - and in the view of the author also a rather leisurely - *first contact* with the order completion method for solving PDEs, there is not available anything published which would be better than [21, pp. VII-X, 1-293].

In this regard, the *theoretical core* can be found at [21, pp. VII-X, 1-158, 237-293], while certain nontrivial *applications* are at [21, pp. 159-183, 210-236].

Also *extensions* of a surprising generality can be found at [21, pp. 184-209], as far as classes of equations far beyond PDEs can be solved by the order completion method, and in fact, *necessary and sufficient* conditions can explicitly be found for their solvability.

Brief Sketch of the Book

The book presents a recently introduced, [21], *unified* and *general*, that is, *type independent* method both for the *existence* and *regularity* of solutions for very large classes of *nonlinear systems* of PDEs, with possibly associated initial and/or boundary value problems.

The *existence* of solutions is obtained on the whole of the domains on which the PDEs are defined.

The minimal blanket *regularity* of all solutions obtained is that they are always at least Hausdorff-continuous on the whole of the mentioned domains.

As fas as such a *universal regularity* goes, there is no longer any need in dealing with various generalized solutions, be they Schwartz distributions, elements of Sobolev spaces, or for that matter, in any of the nonlinear theories of generalized functions given by the infinitely many possible differential algebras of generalized functions listed under 46F30 by the American Mathematical Society.
Consequently, methods of functional analysis are not needed in order to obtain the mentioned results. It is obvious, however, that methods of functional analysis, as well as any other possible methods, are welcome in order to obtain - whenever of interest - additional solutions, as well as stronger regularity and other properties for solutions whose existence was prior proved by any possible methods, including the method presented here.

In short, the method presented here brings the overall problem of solution of PDEs to a similar level of *generality* with that achieved for ODEs, back in 1894, by Charles Emile Picard, in his Comptes Rendu Acad. Sci. Paris paper, where the Cauchy problem associated with general nonlinear systems of ODEs was solved by the method of successive approximations.

And now, about the method in this book, see details in [21,22,1-3,47-56,58,64-66].

Amusingly, and by the way of Picard, it is also a method of successive approximations ...

The difference, however, is that the approximations are not constructed as by Picard, namely, as a usual sequence of iterations of a certain operator.

Instead, the approximations follow the way Dedekind cuts of rational numbers can approximate irrational numbers.

Here however, instead of the natural linear or total order which exists on the set $\mathbb{Q}$ of rational numbers, and which is used to define the Dedekind cuts, we start with the set $\mathcal{C}^\infty(\Omega)$ of smooth functions on the given arbitrary open domains $\Omega \subseteq \mathbb{R}^n$ on which the PDEs under consideration are defined. And on such sets, we consider the natural *partial* order

$$ f \leq g \quad \Leftrightarrow \quad (\, \forall\, x \in \Omega \,:\, f(x) \leq g(x)\,) $$

for $f, g \in \mathcal{C}^\infty(\Omega)$.

Based on that partial order, as well as on a 1937 theorem of MacNeille, [14,13], one can define Dedekind cuts, and construct the *Dedekind order completion* of $\mathcal{C}^\infty(\Omega)$, as well as of certain suitable partially ordered spaces related to it.

And clearly therefore, this is also a kind of approximation method ...

The book is structured as follows.

Part I : The Main Ideas and Facts

Section 0 recalls the long known issue of "paradigm rigidity" in science, [10,11], which is a consequence of the fact that one of the significant limitations of human intelligence is what may be called as "competence rigidity".

Regarding PDEs, this "competence rigidity" has led to a seemingly unknown sort of "scandal" ...

Namely, the nearly exclusive modern methods of study of PDEs have all belonged - and they still do so - to functional analysis.
At the same time, by far the most unified, general, that is, type inde-

pendent result regarding the existence, uniqueness and regularity of solutions was obtained in the well known classical Cauchy-Kovalevskaia theorem for arbitrary nonlinear systems of analytic PDEs.

On the other hand, the fact is, and it still remains so that, in spite of all of the developments of functional analysis, these developments did not manage to improve even marginally on the Cauchy-Kovalevskaia theorem, when the results of this theorem are considered in their own original classical terms.

Section 1 starts by mentioning some of the recent views of so called "leading" mathematicians regarding PDEs, views according to which it is claimed that - very much unlike in the case of ODEs, and in particular of the mentioned 1894 general result of Picard - one simply *cannot* ever expect any similarly general results on PDEs.

Further, and in sharp *contradistinction* to the above views, a brief mention in made of no less than *two* general solution methods for very large classes of nonlinear systems of PDEs which were introduced in [21,22], and then independently and not much later in [19].

Section 2 presents the precise mathematical definition of the very general class of nonlinear systems of PDEs to which the order completion method introduced in [21,22] applies.

Section 3 returns briefly to further historical details which help placing in context the order completion method introduced in [21,22].

Section 4 mentions a few relevant aspects of the nonlinear algebraic theory of generalized functions listed under 46F30 by the American Mathematical Society, a theory which has earlier obtained certain new results in solving nonlinear PDEs.

Section 5 presents the very basic solution results for PDEs obtained for the first time in literature by the order completion method introduced in [21,22].

Section 6 points to several comparisons in solving PDEs by the usual functional analytic methods, and on the other hand, by the order completion method introduced in [21,22].

Section 8 recalls the issue of solving most general equations, starting with the equation $x^2 = 2$ which confused Pythagoras two and a half millennia ago. Then it presents the most simple classical ways for the *extension* of the original setup of equations, a setup which - unlike the original one - may offer the existence of solutions. Such solutions, of course, appear to be nothing short of *generalized solutions* in terms of the initial setup. For instance, in the case of Pythagoras, the solution $x = \sqrt{2}$ did not belong to the initial setup which was restricted to the rational numbers $\mathbb{Q}$, and which setup Pythagoras simply did *not* want even to think of extending, due to his own fundamental philosophical reasons, namely that - in moder terms - all numbers were supposed to be constructible by finite numbers of arithmetic operations from the natural numbers $\mathbb{N}$, thus, the could only be rational numbers in $\mathbb{Q}$. Thus in short, Pythagoras was ... against ... generalized solutions ...
Now, since then, we have developed three - and by now classical - way to extend the original setup of equations, in case that setup turns out not to contain solutions. And these three ways are variants constructed by algebraic, order completion, or topological completion methods.
Amusingly, the *order completion* method of "Dedekind cuts", used to construct $\mathbb{R}$ from $\mathbb{Q}$ has so far never been tried in order to solve PDEs. Well, the **novelty** in [21,22] is precisely to use the method of "Dedekind cuts" for solving very large classes of nonlinear systems of PDEs with possibly associated initial and/or boundary value problems.

Section 8 presents the most simple and general reasons why solving PDEs - even in the linear constant coefficient case - leads often to the need to *extend* the original setup in which those equations have been formulated.
In this way, this section is a kind of ... elevation ... of the ancient conundrum of Pythagoras to the level of PDEs ...

Section 9 presents one of the most surprising *major* advantages of the order completion method when solving PDEs. Namely, it explains why this method does simply and for ever *eliminate* that most troublesome dichotomy between the "linear" and "nonlinear" PDEs, when it comes to their solution.

Section 10 recalls what should be of a far deeper and wider concern among mathematicians, namely, the ... so far hidden and unappreciated power of methods based on partial order ...

Indeed, the classical 1936 Freudenthal Theorem is recalled, a theorem which is formulated exclusively in terms of partial order, and then it is proved in the same terms.

And yet, it has as rather trivial particular cases, that is, consequences, such fundamental results in no less than *three* important branches of mathematics, like :

- the celebrated spectral representation theorem for normal operators in Hilbert spaces,

- the highly nontrivial Radon-Nikodym theorem in measure theory, and

- the Poisson formula for harmonic functions in an open circle.

Section 11 tries to draw certain general conclusions.

Section 12 Appendix : Trying to Ease ... the Pains of ... Paradigm Rigidity ... is a likely to be ... somewhat controversial ... detour into the ... inevitable fundamental phenomenon of "paradigm rigidity", which in fact is merely a ... polite ... way to talk about ... paradigm hostility ...

References

Part II : Six Papers on the Order Completion Method

1. A Brief Announcement about the Increased Blanket Regularity of Solutions, and the Definition of Their Property

2. More Detailed Presentation of the Stronger Blanket Regularity Property of Solutions

3. Certain Details on Solving Large Classes of Nonlinear Systems of PDEs

4. A Few General Results on Partial Orders

5. Solving Arbitrary Equations by Order Completion : Necessary and Sufficient Conditions for the Existence of Solutions

6. Further Details on Solving PDEs by Order Completion

Part III : A Few Practical Suggestions ...

What to Read First, and How to Try to Read it ...

0. A glimpse into History related to Science and Paradigms, and the inescapable ... Competence Rigidity ...

Nowadays history may be of interest only if it is about some amusing anecdote ...
Here therefore, hopefully, may be such one, related to me by a close friend.
When my friend, whom we shall call X, was young, about six decades earlier, through some family of one of his friends, and the family of one of the friends of that family, got in close contact with the world chess champion at that time, whom we shall call Y. And such a strange chain of connections need not much surprise, since it only involved about five or six steps at most, while it has been known for sometime by now that a chain of no more than six steps does exist between just about every two human beings ...

Well, the story is that one of the top chess players with whom Y liked to train often, kept suggesting that it may be useful to play chess also under slightly modified rules. And they did so for a while quite frequently ...
What happened was the realization that, no matter how little they would change the rules of chess, the proficiency of all of them - including of the world chess champion X - would instantly drop quite significantly, and it would not be easy at all to come back to a level of playing anywhere near to that which they had in the usual, unmodified game of chess ...

As it happens, that phenomenon of - shall we call it *competence rigidity* - is highly typical for us humans, and expresses one of the *major limitations* of our intelligence as such ...

Of course, such a rigidity need not manifest itself in realms with less rigorous rules than chess, for instance. However, by the same token, it does manifest itself equally in hard sciences as well, among them, needless to say, in mathematics ...

This phenomenon - as far as science is concerned - has been brought into a well documented and argued focus of general awareness under the name of *paradigm*, by Thomas Kuhn, in his celebrated 1962 book "The Structure of Scientific Revolutions", [10,11]. And the regrettable phenomenon related to it is again the mentioned *rigidity* ...

Thus altogether, and regarding science, we may be speaking about *paradigm rigidity* as being one of the most obvious and unfortunate limitations of human intelligence ...

All in all, and when spreading unfortunately also into various human ventures beyond the realms of science, one may perhaps call it "competence rigidity" ...

And the term "rigidity" may indeed be appropriate, and more so than the term "resistance to change" for instance. Indeed, "rigidity" includes "resistance to change" as well. However, unfortunately, it includes other *negative* manifestations as well, such as, active and manifest, and why not, devious and hidden hostility, among others ...

Certainly, "competence rigidity" also comes with considerable discomfort on the part of those who happen to manifest it ...

There is however, a second kind of rather general limiting effect in human intelligence, one that inevitably gets manifested in science as well.

And being here in this book concerned about mathematics, we shall focus the ways of that limiting effect to the mathematical way of thinking alone.

Well, in mathematics, like in other intellectual ventures, there is a habit to try to classify the diverse ways of thinking exhibited by different mathematicians. One of the simple such classification, for instance, is that in the following two types :

- problem solvers,

versus

- theory builders.

Not long ago, another somewhat more picturesque related classification was resuscitated by the Princeton celebrity physicist, Freeman Dyson, born in 1923, who published it in the item "Birds and Frogs" in the Notices of the AMS, 2009, February, Volume 56, Number 2, pages 212-223, where the "birds" may broadly correspond to "theory builders", while the "frogs" to "problem solvers" ...

Amusingly, Dyson considers himself to be a "frog" ...
But then, who are we to argue with that ?

One may still object that Dyson missed noticing some mere ... "worms" ... who, as a subclass of "frogs", like to get lost in technically highly complicated details ...

The issue however is with the fact that the mentioned major human limitation of what we called competence rigidity is quite likely yet more manifest with "frogs" and "worms", than with "birds" ...
Not that the all the "birds" would significantly be free from that limitation ...
After all, most of the "birds" would anyhow manage to build only one single theory in all of their life, and then, they may be seriously tempted to limit themselves to it for ever after ...

In view of the above, and rather as so often, the optimal situation would be when one has a good amount of all the three features, namely, of "birds", "frogs", and why not, even "worms" ...
And why not, all of that topped by a minimal ... competence rigidity ...

Regarding the "bird", "frog" and "worm" views, certain details can be found in the Appendix.

Now let us focus further somewhat more, namely, from mathematics in general, to PDEs which are supposed to be the subject of this book.

Well, the way PDEs are treated nowadays, and have been treated for nearly a century by now, and certainly since Hilbert in the 1920s, Sobolev in the 1930s, not to mention the Schwartz distributions introduced in the late 1940s, the mathematics involved is mostly functional analysis.

In this regard, the "good news" is represented by the extraordinary amount and variety of major results obtained.

On the other hand, the "bad news" - however not much considered in PDE circles - is in the fact that "analysts" - and among them of course functional analysts - are quite naturally to be found more in the "frog" and "worm" classes, than in the "bird" one ...

Furthermore, one cannot ... accuse ... analysts or functional analysis of having little, or in fact not any, of that regrettable ... competence rigidity ...
Indeed, ... deficiency ... in competence rigidity may be quite high among analysts or functional analysts, not to mention mathematicians in many other branches of mathematics ...

One fact bordering on the ... anecdotic ... in this regard is quite telling, indeed :

> At PDEs conferences, of which there are quite many, when two participants meet one another the first time, one of the first questions they ask about one another is whether one is "elliptic", "parabolic", or "hyperbolic", that is, whether one specializes in one of the respective types of PDEs ?

An effect of the above is that, among PDE specialists, it is simply *inconceivable* that the near monopoly of functional analytic methods - lasting by now for nearly a century - could ever be set aside in the foreseeable future ...

And with this most firmly entrenched view, the following most obvious and questionable situation never seems to concern anybody at all :

1) The well known classical Cauchy-Kovalevskaia theorem on the existence, uniqueness and regularity of solutions for arbitrary nonlinear systems of analytic PDEs was obtained in 1874.

2) The similarly general type independent result for nonlinear systems of ODEs was obtained by Picard in 1894, that is, two decades later.

3) The Cauchy-Kovalevskaia theorem did not use any functional analysis, since that branch of mathematics simply did not exist at the time.

4) All of the subsequently developed functional analysis has ever since been unable to improve in the least on the results of the Cauchy-Kovalevskaia theorem, when these results are considered in their own original classical terms.

And that situation hardly seems to be in the awareness of anybody among those involved nowadays in PDEs, although it may easily constitute nothing short of a ... "scandal" ...

Well, as mentioned, ever since the 1990s, there have been introduced no less than *two* unified, general, type independent methods for solving large classes of systems of nonlinear PDEs.
The first method, presented in [21,22], is in fact considerably more general than the second one in [19], regarding the nonlinear PDEs to which it applies.
In addition, the first method - unlike the second one - does *not* use functional analysis, and instead, it uses an *order completion* method.

Here it may be useful to point out from the beginning that the order completion method has certain considerable *advantages* over any possible functional analytic method.

Indeed, among others, the concept of order is so general that it simply *cannot* distinguish between linear, and on the other hand, nonlinear entities, and in particular, PDEs.
Consequently, for the order completion method it is equally difficult - or for that matter, easy - to solve linear or nonlinear PDEs.

On the other hand, functional analysis involves vector spaces. Thus one does *inevitably* end up with the immensely sharp *discrimination* between linear and nonlinear entities, and specifically, linear, versus, nonlinear PDEs. And one of the essential features of that discrimination is that dealing with the nonlinear PDEs is incomparably more *difficult*.
Yet at the same time, the class of nonlinear PDEs is incomparably

larger, than that of the linear ones. Not to mention that, added to all of that, comes the fact that in modern science the only basic linear PDE of interest is the Schrödinger equation in the usual quantum mechanics.

Indeed, all the other basic PDEs considered in various sciences and applications in our days are nonlinear ...

1. A Sample of Customary Perception

The 2004 edition of the Springer Universitext book "Lectures on PDEs", by V I Arnold, [4], starts up front and right on page 1, with the statement :

> "In contrast to ordinary differential equations, there is *no unified theory* of partial differential equations. Some equations have their own theories, while others have no theory at all. The reason for this complexity is a more complicated geometry ..." (italics added)

The 1998 edition of the book "Partial Differential Equations" by L C Evans, [8], starts his Examples on page 3, with the statement :

> "There is no general theory known concerning the solvability of all partial differential equations. Such a theory is *extremely unlikely* to exist, given the rich variety of physical, geometric, and probabilistic phenomena which can be modeled by PDE. Instead, research focuses on various particular partial differential equations ..." (italics added)

And yet, in 1994, in [21], see MR 95k:35002, precisely such a unified, general, that is, type independent theory of existence and blanket minimal regularity of solutions for very large classes of nonlinear PDEs was published. For latest developments, see [1-3,47-56,58,64-66].

In the sequel, we present the main ideas and motivations which underlie the order completion method. The detailed mathematical developments can be found in the references mentioned above.

It is on occasion worth recalling that we all do mathematics based on certain underlying ideas and motivations. What happens is that we may hold to them for longer, and do so long enough, so that many of them may become rather automatic. And once that happens, we do no longer - and in fact, can no longer - review them, and do so at least now and then.
This is, then, how perceptions are established, and we end up being subjected to them.

Here an attempt is made to go beyond such perceptions in the realms of solving PDEs. And since perceptions are inevitably formulated in some kind of "meta-language" - in this case "meta" with respect to the usual mathematical texts - much of what follows has to go along with that.
Yes, all this is but an unavoidable attempt to ... avoid ... the regrettable effects of ... paradigm rigidity ..., and in general, of ... competence rigidity ...

The fact is - and remains - that, recently, two rather different unified and general, that is, type independent solution methods have been developed for very large classes of linear and nonlinear systems of PDEs with possibly associated initial and/or boundary conditions. The first one of them, and the more general one, [21,22,1-3,47-56,58,64-66], is based on a new idea in the realms of solving PDEs, namely, the *Dedekind order completion* of suitable spaces of smooth functions, while the somewhat latter one, [19], is using standard functional analytic methods.

Contrary to widespread perceptions, it thus proves to be possible to implement no less than two powerful solution methods for a very large variety of linear and nonlinear PDEs. These two methods are *type independent* in the sense that they are *no longer* dependent on specifics of one or another of the countless particular types of PDEs.
In fact, the essence of both methods is that, each in its own way is able to solve far more general equations than PDEs. And it is precisely in this lifting to a higher level of generality, a level considerably beyond PDEs, that the two methods attain their respective type independent power.

These two solution methods have somewhat different reach. The method in [21,22,1-3,47-56,58,64-66] can deal with considerably more general equations, and among them linear and nonlinear systems of PDEs with possibly associated initial and/or boundary conditions. The method in [19] does in fact deliver not only the existence of solutions, but also efficient numerical methods for approximating them.

In view of the above, it is clear that in proving the *existence* of solutions with the mentioned *regularity* properties, the order completion method can *dispense* to a good extent with all sorts of spaces of generalized functions, distributions, hyperfunctions, and so on, and instead, focus on various classes of usual measurable functions or Hausdorff continuous ones which are defined on usual Euclidean domains.

Certainly, the use of generalized solutions need not be avoided completely. However, solution methods based on them are clearly far less powerful, especially with respect to their existence, than the order completion method. As for the regularity of solutions obtained by the order completion method, it gives from the beginning solutions which are more regular than by far most generalized functions, since they can be assimilated with usual measurable functions or Hausdorff continuous ones on Euclidean domains.

2. The Very Large Class of Nonlinear Systems of PDEs Solved

In [21,22] it was show how to obtain solutions U for all systems of nonlinear PDEs with possibly associated initial and/or boundary value problems, where the equations involved are of the very general form

(2.1) $\quad F(x, U(x), \ldots, D_x^p U(x), \ldots) = f(x), \ x \in \Omega \subseteq \mathbb{R}^n, \ |p| \leq m$

Here F is any function *jointly continuous* in all its arguments, the right hand term f can belong to a class of *discontinuous* functions, the order $m \in \mathbb{N}$ is given arbitrary, while the domain Ω can be any bounded or unbounded open set in $\mathbb{R}^n$.

In fact, even the functions F defining the nonlinear partial differential operators in the left hand terms of (2.1) can have certain types of *discontinuities*.

Here indeed, in view of (2.1), one can note the *unprecedented* generality, type independence, or universality of the corresponding result both on the *existence* and the *regularity* of solutions given in [21,22] for systems of nonlinear PDEs constituted from equations of the above form (2.1).

Indeed, regarding the *existence* of solution, the generality of the PDEs in (2.1) is self-evident.

As for the *regularity* of the solutions obtained, one can note the following.

The solutions U obtained in [21,22] can be assimilated with *usual measurable functions* on the respective domains Ω.
Not much later, in [1-3,47-56,58,64-66], it was shown that such solutions are in fact always considerably more regular, being in fact Hausdorff continuous functions on the whole of the domains Ω of the respective PDEs.

Recently, in [47-56] this general, type independent, universal, or blanket regularity result was further improved as it was shown that the solutions U can be associated with usual smooth functions on the domains Ω of the PDEs concerned, provided that similar smoothness conditions are assumed on the functions F and f in the PDEs (2.1).

Here it is important to note that *Hausdorff continuous* functions are not much unlike usual real valued continuous functions, [1-3]. Indeed, on suitable *dense* subsets of their domains of definition, Hausdorff continuous functions have as values real numbers, and are completely *determined* by such values. On the rest of their domains of definition, Hausdorff continuous functions can have values given by bounded or unbounded closed intervals of real numbers. Also, every real valued function which is continuous in the usual sense will be Hausdorff continuous as well.

One of the major advantages of the order completion method is that it *eliminates* the rather mercilessly difficult to handle dichotomy between linear, and on the other hand, nonlinear PDEs, treating both cases with equal ease. Indeed, the dichotomy between linear and nonlinear follows from the vector space structure of the spaces of functions on which the partial differential operators act. In this way, this dichotomy is of an algebraic nature. On the other hand, partial orders are more basic mathematical structures than algebra, and as such, they do not, and simply cannot, differentiate between linear and nonlinear entities.

Clearly, functional analytic methods, which rely not only on topological but also algebraic structures, cannot exhibit such a performance, since they are bound to discriminate between linear and nonlinear entities, and in particular, equations, see details in section 9.

3. A Short History of Difficulties in Solving Linear and Nonlinear PDEs

The first unified and general, that is, *type independent* existence result for solutions of rather arbitrary nonlinear systems of PDEs was obtained in 1874, when upon the suggestion of K Weierstrass, Sophia Kovalevskaia gave a rigorous proof for an earlier theorem of Cauchy, published in 1821, in his Course d'Analyse. This result although completely general as far as the type independent nonlinearities involved ar concerned, assumes however, that in the systems of PDEs of the form (2.1) both F and f are analytic. In addition one also assumes initial value problems on non-characteristic analytic hyper-surfaces, while boundary value problems are not treated by the respective Cauchy-Kovalevskaia theorem.

However, in such a highly particular situation concerning the regularity of the PDEs and the initial data involved, the solutions obtained are proved to exist always, and also to be unique and analytic.

The problem with that classical *existence, uniqueness* and *regularity* result is that, typically for nonlinear PDEs, such analytic solutions do not - and in general, simply cannot - exist globally on the whole of the domain of the respective PDEs, but only in certain neghbourhoods of

the analytic hyper-surfaces on which the initial values are given.
This limitation is, therefore, not due to the specific method of proof of Kovalevskaia.

Indeed, it already happens in such most simple nonlinear PDEs as the well known shock wave equation

$$U_t(t,x) + U(t,x)U_x(t,x) = 0, \quad t \geq 0, \ x \in \mathbb{R}$$

$$U(0,x) = u(x), \quad x \in \mathbb{R}$$

where precisely the non-global solutions are of interest, since among them are the so called shock wave solutions. And such shock waves do happen always when, for instance, the initial value u, no matter how smooth, including possibly analytic, happens to be decreasing on no matter how small subinterval in $\mathbb{R}$.

And clearly, the above shock wave equation is covered by the classical Cauchy-Kovalevskaia theorem, whenever the initial value u is analytic.

Here however it is important to note that the failure of the existence of global analytic solutions is but a part of a far more general phenomenon, since even linear, let alone nonlinear PDEs may fail to have smooth, or even merely classical solutions, even in the case of solutions of major applicative interest. After all, even linear constant coefficient PDEs have non-classical solutions of particular interest, such as those given by Green functions.

What may be interesting, and also worthwhile to note with respect to the mentioned Cauchy-Kovalevskaia theorem, are the following three facts :

> 1) The rigorous proof by Kovalevskaia of that theorem on solutions of general nonlinear systems of PDEs predates by two decades the corresponding general theorem on solving systems of nonlinear ODEs defined by continuous functions. Indeed, the existence of solutions for such ODEs was given by Charles Emile Picard in his 1894 Comptes Rendu Acad. Sci. Paris paper, where the associated Cauchy problem was solved by the method of successive approximations.

2) The only so called "hard" mathematics used in the proof of the Cauchy-Kovalevskaia theorem is the elementary formula for the summation of a convergent geometric progression. The rest of the proof is but a succession of rather elementary, even if quite involved, estimates of terms in power series. In this way, the proof of the Cauchy-Kovalevskaia theorem does *not* involve any methods of functional analysis. And certainly it could *not* involve such methods at the time in the 1870s, methods which simply were inexistent, when that proof was given. On the other hand, the proof of the corresponding general existence result for solutions of nonlinear systems of ODEs does involve a fixed point argument in suitable spaces of functions which are complete in their respective topologies.

3) The result in the Cauchy-Kovalevskaia theorem - when considered on its own original terms of type independent nonlinear generality - could *not* so far be improved in those very terms, regardless of all the advances in functional analysis of the last more than a century.

The only more important improvement of the classical result in the Cauchy-Kovalevskaia theorem was obtained in 1985, without however using functional analytic methods, see section 4. And this improvement delivered a *global* existence result for solutions, as well as a regularity of such solutions which guarantees their *analyticity* on the whole of the domains of definition of the respective PDEs, except possibly for closed, nowhere dense subsets which can be chosen so as to have zero Lebesgue measures.

Indeed, when it comes to *type independent nonlinear generality*, the functional analytic methods used in solving PDEs could bring about improvements - and often quite dramatic ones - only in a variety of far more particular cases, than the type independent nonlinear generality dealt with in the classical Cauchy-Kovalevskaia theorem.

In this way, in spite of more than one century of functional analysis, the classical Cauchy-Kovalevskaia theorem still remains a *maximal* result, except for its extension mentioned in section 4, which again, does not use functional analysis.

In the early 1950s, soon after the introduction of the linear theory of distributions by L Schwartz, it was proved independently by Malgrange, [15], and Ehrepreis, [7], that in case of a single PDE of the form (2.1), if the left hand term F is linear and with constant coefficients, while f is the Dirac delta distribution, then (2.1) always has a global, so called, fundamental solution given by a suitable Schwartz distribution.

This rather general, that is, type independent linear result appeared to suggest that a similar result could be obtained in the more general case when F in (2.1) is linear and with smooth coefficients. L Schwartz himself is known to have conjectured such a generalization, and furthermore, as it appears, he suggested it at the time to Francois Treves as a subject for his doctoral thesis, as mentioned in [46].

However, in 1957, Hans Lewy, [12], showed that the rather simple linear first order PDE in three space variables and with first degree polynomial coefficients

$$(3.1) \quad (D_x + iD_y - 2(x+y)D_z)\, U(x, y, z) \;=\; f(x, y, z),$$

with $(x, y, z) \in \mathbb{R}^3$, does *not* have any Schwartz distribution solutions in any neighbourhood of any point in $\mathbb{R}^3$, for a large class of smooth right terms f.
Furthermore, in 1967, Shapiro, see it mentioned in [30,31], gave a similar example of a smooth linear PDE which does not have solutions in Sato's hyperfunctions.

In the early 1960s, L Hörmander gave certain *necessary* conditions for the solvability in distributions of arbitrary linear smooth coefficient PDEs, see [30,31].

All such linear results were appearing, however, in the shadow of the 1954 so called "impossibility result" of Schwartz, [45], which continued to be misunderstood for a longer time, namely, by being seen as a proof for the "impossibility to multiply distributions", as claimed among others by H$'$ormander in 1976, [28-31], that is, more than two decades after the publication of the mentioned Schwartz paper.

Certainly, such misinterpretations have delayed the development and acceptance of sufficiently general and systematic *nonlinear* theories of generalized functions, such as are, since 1999, listed under 46F30 in the Mathematics Subject Classification of the American Mathematical Society. Extensive details - with an *overall* analysis - can be found in [40], while its several decades earlier beginnings are, among others, in [24-29,6,30-39,16-18,42-44,57,58,61-63,67,70,71,74,75].

Regarding the perception that nontrivial general, type independent results are just about impossible to obtain related to PDEs, it is worth noting that the mentioned Malgrange-Ehrenpreis result on fundamental solutions is in fact precisely such a nontrivial general and type independent existence result within the range of all linear and constant coefficient PDEs.
The necessary condition for the existence of distributions solutions given by Hörmander is also a nontrivial general and type independent result, this time within the much larger class of all linear smooth coefficient PDEs.

In this regards, of course, the classical Cauchy-Kovalevskaia theorem is the most impressive *classical* general and type independent nonlinear result regarding existence, regularity and uniqueness of solutions, although it is obtained without any kind of functional analysis.

4. Nonlinear Algebraic Theory of Generalized Functions

This nonlinear theory - see 46F30 in the AMS Subject Classification 2010 - was started in the 1960s, [24-40,42-44,57,58,61-63,67,70,71,74,75], and it is based on the construction of all possible - and in fact, not less than *infinitely* many - differential algebras of generalized functions which contain the Schwartz distributions.
That theory has managed to come quite near to solving the Lewy impossibility (3.1), in the case of linear PDEs with smooth coefficients. Yet it did not solve it completely, although it obtained generalized

function solutions for large classes of linear and nonlinear PDEs. As an example, and as mentioned, back in 1985, it obtained the first *global* existence result for the general nonlinear PDEs in the classical Cauchy-Kovalevskaia theorem. And the respective global solutions are analytic on the whole of the domain of the PDEs, except for certain closed and nowhere dense subsets, which can be chosen to have zero Lebesgue measure, see [30, pp. 259-266], [31, pp. 101-122], [32,39,40].

5. The Order Completion Method

Surprisingly, the order completion method in solving general nonlinear systems of PDEs of the form (2.1) is based on certain very simple, even if less than usual, approximation properties, see [21, pp. 12-20]. To give here an idea about the way the order completion method works, we mention some of these approximations here in the case of one single nonlinear PDE of the form (2.1).

Let us denote by $T(x, D)$ the left term in (2.1), then we have the basic approximation property :

Lemma 5.1

$\forall \ x_0 \in \Omega, \ \epsilon > 0 \ :$

$\exists \ \delta > 0, \ P$ polynomial in $x \in \mathbb{R}^n \ :$

$$||x - x_0|| \leq \delta \implies f(x) - \epsilon \leq T(x, D)P(x) \leq f(x)$$

□

Consequently, we obtain :

Proposition 5.1

$\forall \ \epsilon > 0 \ :$

$\exists \ \Gamma_\epsilon \subset \Omega$ closed, nowhere dense in $\Omega, \ U_\epsilon \in C^\infty(\Omega) \ :$

$$f - \epsilon \leq T(x, D)P \leq f \text{ on } \Omega \setminus \Gamma_\epsilon$$

Furthermore, one can also assume that the Lebesgue measure of Γ_ϵ is zero, namely

$$mes\ (\Gamma_\epsilon)\ =\ 0$$

$\square$

Let us now note that, see [1-3]

$$\mathcal{C}^0(\Omega) \subset \mathbb{H}(\Omega)$$

and the set $\mathbb{H}(\Omega)$ of *Hausdorff continuous* functions on Ω is Dedekind order complete.
Consequently, we obtain the following basic result on the *existence* and *regularity* of solutions for nonlinear PDEs of the form (2.1) :

Theorem 5.1

$$T(x,D)U(x)\ =\ f(x),\quad x \in \Omega$$

has solutions U which can be assimilated with Hausdorff continuous functions, for a class of discontinuous functions f on Ω, class which contains the continuous functions on Ω.

$\square$

We give here some more details related to Theorem 5.1 above. In view of Proposition 5.1, we shall be interested in spaces of piecewise smooth functions given by

$$(5.1)\quad \mathcal{C}^l_{nd}(\Omega)\ =\ \left\{ u\ \left|\ \begin{array}{l} \exists\ \Gamma \subset \Omega\ \text{closed, nowhere dense }:\\ *)\ u : \Omega \setminus \Gamma \to \mathbb{R}\\ **)\ u \in C^l(\Omega \setminus \Gamma) \end{array} \right. \right\}$$

where $l \in \mathbb{N}$. It is easy to see that we have the inclusions

$$(5.2)\quad T(x,D)\,\mathcal{C}^m_{nd}(\Omega) \subseteq \mathcal{C}^0_{nd}(\Omega) \subset \mathbb{H}(\Omega)$$

In this way, we obtain the following more precise formulation of the result in Theorem 5.1 on the existence and regularity of solutions :

Theorem 5.1*

(5.3) $\quad T(x, D)^\# \, (\mathcal{C}^m_{nd}(\Omega))^\#_T \;=\; (\mathcal{C}^0_{nd}(\Omega))^\# \;\subset\; \mathbb{H}(\Omega)$

$\square$

Here $(\mathcal{C}^m_{nd}(\Omega))^\#_T$ and $(\mathcal{C}^0_{nd}(\Omega))^\#$ are Dedekind order completions of $\mathcal{C}^m_{nd}(\Omega)$ and $\mathcal{C}^0_{nd}(\Omega)$, respectively, when these latter two spaces are considered with suitable partial orders.

The respective partial order on $\mathcal{C}^m_{nd}(\Omega)$ may depend on the nonlinear partial differential operator $T(x, D)$ in (5.2), while the partial order on $\mathcal{C}^0_{nd}(\Omega)$ is the natural point-wise one, at the points where two functions compared are both continuous.
The operator $T(x, D)^\#$ is a natural extension of the nonlinear partial differential operator $T(x, D)$ in (5.2) to the mentioned Dedekind order completions.

The meaning of (5.3) is twofold :

- for every right hand term $f \in (\mathcal{C}^0_{nd}(\Omega))^\#$ in (2.1), there exists a solution $U \in (\mathcal{C}^m_{nd}(\Omega))^\#_T$, and the set $(\mathcal{C}^0_{nd}(\Omega))^\#$ contains many discontinuous functions beyond those piecewise discontinuous ones, see [21],

- the solutions U can be assimilated with Hausdorff continuous functions on Ω, see [1-3].

6. Comparison with Methods in Functional Analysis

The order completion method is a powerful *alternative* and *complement* to the usual functional analytic ones, when solving linear or nonlinear PDEs. Details in this regard are presented in [21, chap. 12]. Certainly, the order completion method is *not* meant to replace the functional analytic ones, the latter being useful in obtaining stronger results in a large variety of particular PDEs.

Here, we shall only mention the following. Functional analytic methods in solving PDEs are based on the *topological completion* of uniform spaces, such a normed or locally convex vector spaces of suitably chosen functions. In this respect, the comparative advantages of the order completion method can shortly be formulated as follows :

- unlike the functional analytic methods, which are geared more naturally to the solution of linear PDEs, the order completion method performs equally well in the case of both linear and nonlinear PDEs, see section 9 below,

- unlike the functional analytic methods, which face considerable difficulties when dealing with initial, and especially, boundary value problems, the order completion method performs without any significant additional troubles in such situations,

- the order completion method gives solutions which can be assimilated with usual measurable, or even Hausdorff continuous functions, and thus the solutions obtained are not merely distributions, generalized functions or hyper-functions.

As an illustration of the comparative situation regarding these two methods let us consider on a bounded Euclidean domain Ω, which has a smooth boundary $\partial \Omega$, the following well known linear boundary value problem

$$(6.1) \quad \begin{array}{l} \Delta U(x) = f(x), \quad x \in \Omega \\ U = 0 \text{ on } \partial \Omega \end{array}$$

As is well known, for every given $f \in C^\infty(\widetilde{\Omega})$, this problem has a unique solution U in the space

$$(6.2) \quad X = \left\{ v \in C^\infty(\widetilde{\Omega}) \mid v = 0 \text{ on } \partial \Omega \right\}$$

It follows that the mapping

35

(6.3) $\quad X \ni v \longmapsto \|\Delta v\|_{\mathcal{L}^2(\Omega)}$

defines a norm on the vector space X. Now let

(6.4) $\quad Y = \mathcal{C}^\infty(\widetilde{\Omega})$

be endowed with the topology induced by $L^2(\Omega)$. Then in view of (6.1) - (6.4), it follows that the mapping

(6.5) $\quad \Delta : X \to Y$

is a uniform continuous linear bijection. Therefore, it can be extended in a unique manner to an isomorphism of Banach spaces

(6.6) $\quad \Delta : \widetilde{X} \to \widetilde{Y} = \mathcal{L}^2(\Omega)$

In this way one has the classical existence and uniqueness result

$$\forall \ f \in \mathcal{L}^2(\Omega) \ :$$

(6.7) $\quad \exists! \ U \in \widetilde{X} \ :$

$$\Delta U = f$$

The power and simplicity - based on linearity and topological completion of uniform spaces - of the above classical existence and uniqueness result is obvious. This power is illustrated by the fact that the set $\widetilde{Y} = \mathcal{L}^2(\Omega)$ in which the right hand terms f in (6.1) can now be chosen is much *larger* than the original $Y = \mathcal{C}^\infty(\widetilde{\Omega})$.

Furthermore, the existence and uniqueness result in (6.7) does not need the a priori knowledge of the structure of the elements $U \in \widetilde{X}$, that is, of the respective generalized solutions. This structure which gives the regularity properties of such solutions can be obtained by a further detailed study of the respective differential operators defining the PDEs under consideration, in this case, the Laplacean Δ. And in the above specific instance, in the respective Sobolev spaces, we obtain

(6.8) $\quad \widetilde{X} = H^2(\Omega) \cap H^1_0(\Omega)$

As seen above, typically for the functional analytic methods, the generalized solutions are obtained in topological completions of vector spaces of usual functions. And such completions, like for instance the various Sobolev spaces, are defined by certain linear partial differential operators which may happen to *depend* on the PDEs under consideration.

In the above example, for instance, the topology on the space X obviously *depends* on the specific PDE in (6.1). Thus the topological completion $\tilde{X}$ in which the generalized solutions U are found according to (6.7), does again *depend* on the respective PDE.

On the other hand, with the method of order completion we are *no longer* looking for generalized solutions, and instead, a *type independent, universal* or *blanket* regularity property is attained, since the solutions obtained can always be assimilated with usual measurable functions, or even with Hausdorff continuous functions.

Similar to the functional analytic methods, however, the order completion method obtains the solutions in spaces which may again be related to the specific nonlinear partial differential operators $T(x, D)$ in the equations of form (2.1).

7. Solving General Equations by Extending their Domains of Definition : the Three Classical Methods

The ancient case of *solving an equation*, which shocked Pythagoras two and a half millennia ago, is given by

(7.1) $\quad x^2 = 2$

This is of the general form

(7.2) $\quad E(x) = c$

where for arbitrary nonvoid sets X and Y, we are given any mapping

(7.3) $\quad E : X \to Y$

together with a specific $c \in Y$, and then we want to find a solution $x \in X$, so that (7.2) holds.

What shocked Pythagoras was that (7.1) could not be solved, if one restricted oneself to $X = \mathbb{Q}$ in (7.3). And it took no less than about two millennia or more, till we could rigorously extend $X = \mathbb{Q}$ to $\tilde{X} = \mathbb{R}$, and thus obtain a well defined solution $x = \pm\sqrt{2}$ of (7.1).

Clearly, the morale of the above conundrum which Pythagoras did not know how to deal with, is in fact most trivial. Namely, any equation (7.2) can only be solved by a solution $x \in E$, for arbitrary $c \in Y$, if and only if the mapping (7.3) happens to be *surjective*.

In this way, ever since, we have the following model lesson in solving equations :

- if one cannot solve (7.2) within the framework of (7.3), then one can try to solve it in the *extended* framework

(7.4) $\quad \tilde{E} : \tilde{X} \to \tilde{Y}$

where $X \subset \tilde{X}$, $Y \subseteq \tilde{Y}$, and $\tilde{E}$ is such that we have the commutative diagram

(7.5)
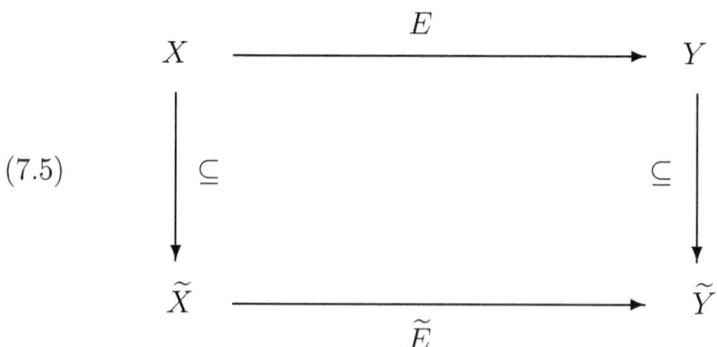

Here however, we face the following problems :

- how to choose or construct $\tilde{X}$, and then how to interpret the new, or so called *generalized* solutions $x \in \tilde{X} \setminus X$, which two questions altogether constitute but the celebrated *regularity* problem regarding the *generalized* solutions obtained,

- how to do the same for $\tilde{Y}$, which nevertheless need not always be done, since we can often stay with Y in (7.4) and only have to extend X to $\tilde{X}$,

- how to define the extension $\tilde{E}$, which often, and typically in the nonlinear case, is not a trivial problem.

Fur further detail, we can now recall that with the equation (7.1) we had to

(7.6) $\quad go\ from \quad \mathbb{Q} \quad to \quad \tilde{\mathbb{Q}} = \mathbb{R}$

On the other hand, with the equation

(7.7) $\quad x^2 + 1 = 0$

we had to

(7.8) $\quad go\ from \quad \mathbb{R} \quad to \quad \mathbb{C}$

However, there is a vast difference between (7.6) and (7.8). Consequently, there is also a vast difference between solving (7.1) and (7.7).

Indeed, we solve (7.7) through the extension (7.8) which is a mere *algebraic adjoining* of an element, in this case, of $i = \sqrt{-1}$ to $\mathbb{R}$.

On the other hand, when solving (7.1), the extension (7.6) can be seen in general as at least *three different*, even if in this particular case, equivalent, constructions, namely, through :

- topology
- algebra
- order.

39

And to be more precise, we have :

- The Cauchy-Bolzano method which is *ring theoretic* plus *topological*, and it is applied to $\mathbb{Q}$, as it obtains $\mathbb{R}$ according to the quotient construction in algebras

$$(7.9) \quad \mathbb{R} = \mathcal{A}/\mathcal{I}$$

where $\mathcal{A} \subset \mathbb{Q}^{\mathbb{N}}$ is the algebra of Cauchy sequences of rational numbers, while $\mathcal{I}$ is the ideal in $\mathcal{A}$ of sequences convergent to zero.

- The method of Dedekind is based on the *order completion* of $\mathbb{Q}$.

However, the Cauchy-Bolzano method itself can be generalized in two directions :

- In the *topological* generalization, the algebraic part can be omitted, and instead, one only uses the *topological completion* of uniform spaces, here of the usual metric space on $\mathbb{Q}$.

- In the *algebraic* generalization it is possible to extract the abstract essence of (7.9), and simply start with a suitable algebra $\mathcal{A} \subseteq \mathbb{Q}^{\mathbb{N}}$, and an appropriate ideal $\mathcal{I}$ in $\mathcal{A}$. Such a construction can indeed be rather abstract, since it need *not* involve any topology on $\mathbb{Q}$ or $\mathbb{R}$. And indeed, such abstract construction happens, for instance, when constructing the nonstandard reals *$\mathbb{R}$, namely

$$^*\mathbb{R} = \mathcal{A}/\mathcal{I}$$

Here indeed, one simply takes $\mathcal{A} = \mathbb{R}^{\mathbb{N}}$, that is, the algebra of all sequences of real numbers, while the ideal $\mathcal{I}$ is defined by any given free ultrafilter on $\mathbb{N}$.

A rather general version of such an abstract approach, which however makes a certain rather limited use of topology, has been introduced and extensively used in the nonlinear algebraic theory of generalized functions under the AMS classification index 46F30, as mentioned in section 4 above.

What is done in the method in 46F30, is to generalize the Cauchy-Bolzano method by retaining its ring theoretic algebraic aspect, while the topological one is weakened to the certain extent of being confined exclusively to the topologies of Euclidean spaces on which the generalized functions are defined.

Now, what is done in the method introduced in [21], and further developed in [1-3,47-56,58,64-66], is the extension of the classical Dedekind order completion method, used in the construction of $\mathbb{R}$ from $\mathbb{Q}$, to suitable spaces of piece-wise smooth functions on the Euclidean domains on which the PDEs considered are defined.

An important fact to note is that both the topological and order completion methods give us the property that

- $\mathbb{Q}$ is *dense* in $\mathbb{R}$

in the respective sense of topology or order. In this way, the elements of the extension of $\mathbb{Q}$, that is, the elements of $\mathbb{R}$, are in the corresponding sense *arbitrarily near* to the elements of the originally extended space $\mathbb{Q}$. Thus the elements in the extension can arbitrarily be *approximated* by elements of the extended space, be it in the sense of topology, or respectively, order.

Furthermore, both through the methods of topology and order, one obtains $\mathbb{R}$ in a *unique* manner, up to a respective algebraic or order isomorphism.

In this way both the topological and order completion methods have the *double* advantage that

- the elements of the extension are not too strange conceptually,

and furthermore

- the elements of the extension are near to elements of the extended space, within arbitrarily small error.

This *density* property remains also valid in the general Dedekind order completion method used in [21,22,1-3,47-56,58,64-66].
Indeed, in (5.3) we have that $\mathcal{C}_{nd}^m(\Omega)$ and $\mathcal{C}_{nd}^0(\Omega)$ are *order dense* in $(\mathcal{C}_{nd}^m(\Omega))_T^\#$ and $(\mathcal{C}_{nd}^0(\Omega))^\#$, respectively.

Connected with the general extension method in (7.3) - (7.5) one can note that, on occasion, the following *convenient* situation may occur : the extended mapping

(7.10) $\quad \widetilde{E} : \widetilde{X} \longrightarrow \widetilde{Y}$

may turn out to be an *isomorphism* of the respective algebraic, topological or order structures used on X and Y, when constructing the corresponding extensions $\widetilde{X}$ and $\widetilde{Y}$. In such a case, and when one has a better understanding of the structure of the elements in $\widetilde{Y}$, one can obtain in addition a *regularity* type result concerning the so called generalized solutions $x \in \widetilde{X} \setminus X$ of the equations (7.2), since such generalized solutions can be *assimilated* - through the isomorphism $\widetilde{E}$ - with the corresponding elements $\widetilde{E}(x) \in \widetilde{Y}$.

This, as detailed briefly below, is precisely in simple terms how that *blanket regularity* property of solutions of PDEs is obtained by the order completion method.

A classical example of such an isomorphism (7.10) happens, for instance, in (6.5), (6.6), when the boundary value problem (6.1) is solved by using well known functional analytic methods.
In that specific instance, however, the suitable further use of functional analytic methods can lead to the *additional* regularity property of generalized solutions in $\widetilde{X}$, as given in (6.8). Nevertheless, the Banach space isomorphism (6.6) - which in that case is but the particular form taken by (7.10) - is in itself already a *first* regularity result about the structure of the elements of $\widetilde{X}$.

The above convenient situation of an isomorphism of type (7.10) can, indeed, appear as well when using the order completion method in solving the very large classes of nonlinear system of PDEs mentioned in section 2. This is the reason why the solutions obtained in [21] could be assimilated with usual measurable functions, while in [1-3], they can be assimilated even with the much *more regular* Hausdorff continuous functions.

More specifically, in (5.3), the extended mappings $T(x,D)^\#$ prove to be *order isomorphisms* between the spaces $\mathcal{C}^m_{nd}(\Omega)^\#_T$ and $\mathcal{C}^0_{nd}(\Omega)^\#$.

This is then, in essence, the reason why the solutions of nonlinear systems of PDEs of the very general form (2.1) could at first be assimilated with usual measurable functions, and can now be assimilated with Hausdorff continuous functions.

8. The Need for Extensions in the Case of Solving PDEs.

Let us now associate with each nonlinear PDE in (2.1) the corresponding nonlinear partial differential operator defined by the left hand side, namely

$$(8.1) \quad T(x,D)U(x) = F(x, U(x), \ldots, D^p_x U(x), \ldots), \quad x \in \Omega$$

Two facts about the nonlinear PDEs in (2.1) and the corresponding nonlinear partial differential operators $T(x,D)$ in (8.1) are important and immediate :

- The operators $T(x,D)$ can *naturally* be seen as acting in the *classical* context, namely

$$(8.2) \quad T(x,D) : \mathcal{C}^m(\Omega) \ni U \longmapsto T(x,D)U \in \mathcal{C}^0(\Omega)$$

while, unfortunately on the other hand

- The mappings in this natural classical context (8.2) are typically *not* surjective. In other words, linear or nonlinear PDEs in (2.1) typically *cannot* be expected to have *classical* global solutions $U \in \mathcal{C}^m(\Omega)$, for arbitrary continuous right hand terms $f \in \mathcal{C}^0(\Omega)$.
  Furthermore, it can often happen that non-classical solutions have a major applicative interest, thus they have to be sought out beyond the classical framework in (8.2).

This is, therefore, how we are led to the *necessity* to consider *generalized solutions* U for PDEs of type (2.1), that is, solutions $U \notin C^m(\Omega)$, which therefore are no longer classical. This means that the natural classical mappings (8.2) must in certain suitable ways be *extended* to *commutative diagrams*

(8.3)
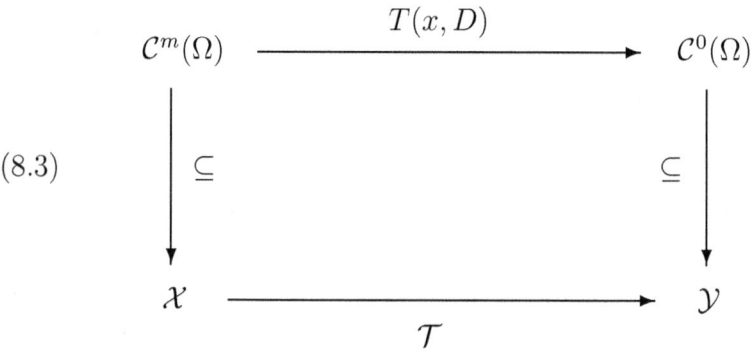

which are expected to have certain kind of *surjectivity* type properties, such as for instance

(8.4) $\quad C^0(\Omega) \subseteq T(\mathcal{X})$

We conclude with a few comments :

- Traditionally, ever since Hilbert and Sobolev, thus starting before WW II, *functional analysis* has been massively used in solving PDEs. And for that purpose, suitable uniform topologies are defined on the domains and ranges of the corresponding partial differential operators $T(x, D)$. These domains and ranges are given by various vector spaces of generalized functions, and in particular, distributions. Thus these operators obtain certain continuity properties. Then the extensions $\mathcal{X}$ and $\mathcal{Y}$ in (8.3) are defined as the *completions* in these uniform topologies of the respective domains and ranges of $T(x, D)$. Finally, the continuity properties of $T(x, D)$ may allow the construction of suitable extensions $\mathcal{T}$ which would give the corresponding versions of the commutative diagrams (8.3), and also satisfy some variant of the surjectivity property (8.4), see for details [21, chap. 12, pp. 237-262].

- Since the 1960s, the *algebraic nonlinear* method in 46F30, see [24–29,6,30-40,42-44,16-18,57,58,61-63,67,70,71,74,75], can alternatively be used especially in the case of *nonlinear* partial differential operators $T(x, D)$. In this respect, a large variety, and in fact, *infinitely many* classes of *differential algebras of generalized functions*, each of them containing the Schwartz distributions, were constructed as the sought after extensions $\mathcal{X}$ and $\mathcal{Y}$ in (8.3).

The most general classes - in fact, no less than *infinitely* many - of such algebras were introduced and used in [24-40,42-44,16-18,57,58,61-63,67,70,71,74,75], starting with the 1960s, [24]. A survey of all such general differential algebras of generalized functions was presented in [40].

Later, in the early 1980s, a particular class of such algebras was introduced in [6], and it has known a certain popularity.
However, due to the specific polynomially type limiting *growth conditions* required in the construction of Colombeau algebras, their use in the study, for instance, of Lie group symmetries of PDEs, or singularities in General Relativity is limited, since in both cases one may have to deal with transformation whose growth can be arbitrary. In this way, such transformations cannot be accommodated within the Colombeau algebras of generalized functions. In fact, in the Colombeau algebras, as well as in other ones which fail to be flabby sheaves, one cannot even formulate - let alone solve - the Global Cauchy-Kovalevskaia Theorem. Similarly, one cannot define globally arbitrary Lie group actions, [74,75].

On the other hand, arbitrary smooth transformations and operations can easily be dealt with in some of the other classes of algebras of generalized functions introduced earlier in [24-40]. Also as mentioned, the Global Cauchy-Kovalevskaia Theorem can be not only formulated, but also solved, and solve globally, which is a first in the literature. Further, arbitrary Lie group actions can be defined globally in such algebras.

In fact, based on such a definition of arbitrary global Lie group actions on generalized functions, in [35] a first time *complete solution* of Hilbert's Fifth Problem was presented.

The severe limitations on dealing with singularities which various vector spaces of distributions or algebras of generalized functions suffer from is closely related to their failure to be *flabby sheaves*, [40]. Among such algebras are the Colombeau algebras which fail to be flabby sheaves due to the mentioned growth conditions which play an essential role in their definition. In particular, Hilbert's Fifth Problem cannot even be formulated, let alone solved, in its full generality in the Colombeau algebras.

Details related to the ability of various vector spaces of distributions and algebras of generalized functions to deal with large classes of singularities are presented in [40]. And it turns out that those vector spaces of distributions and algebras of generalized functions which fail to be flabby sheaves also fail in dealing with large enough classes of singularities.

- The *order completion* method, introduced and developed in 1994, in [21,22], and further developed and improved in [1-3,47-56,58,64-66], constructs the extensions $\mathcal{X}$ and $\mathcal{Y}$ in (8.3) as the *Dedekind order completion* of spaces naturally associated with the partial differential operators $T(x, D)$, and the spaces $\mathcal{C}^m(\Omega)$ and $\mathcal{C}^0(\Omega)$ in (8.2).

Related to the advantages of the order completion method in solving nonlinear PDEs let us mention here in short the following.

Neither the functional analytic, nor the algebraic methods can so far come anywhere near to solve nonlinear PDEs of the generality of those in (2.1), let alone systems of such nonlinear PDEs together with associated initial and/or boundary value problems.

In fact, the functional analytic methods are still subjected to the celebrated 1957 Hans Lewy impossibility which they are nowhere near to manage to overcome, even in the general smooth coefficient linear case.

The extent of this failure within the present functional analytic methods is also illustrated by the necessary condition for the solvability in distributions of linear smooth coefficient PDEs given by Hörmander in the 1960s, a condition which is also sufficient in the case of first order such PDEs, as shown not much later by Nirenberg & Treves, see [30, pp. 37-47] for details. Indeed, ever since that 1960s, this necessary condition could not be improved in the case of second and higher order smooth coefficient linear PDEs by one which would be both necessary and sufficient. And the reason seems to be the considerable complexity involved in the locally convex topological structures which one inevitably encounters when dealing with Schwartz distributions ...

As far as the algebraic method is concerned, it has among others come quite near to the solution of the Lewy impossibility, see [6], and for a short respective account [30, pp. 37-39].
Further powerful results in solving various classes of nonlinear PDEs, not treated so far by the functional analytic method, can be found in [28-33], [6], or [20].

Among such results is the *global* solution of arbitrary analytic nonlinear systems of PDEs, when considered with analytic non-characteristic Cauchy initial values, mentioned in section 4 above. The respective generalized solutions obtained are *analytic* functions, except possibly for *closed* and *nowhere dense* subsets Γ of the domains Ω of definition of the given PDEs. In addition, these subsets Γ can also be chosen to have *zero* Lebesgue measure.

On the other hand, the order completion method introduced and developed in [21,22], and further improved in [1-3,47-56,58,64-66] as far as the *regularity* of solutions is concerned, can not only deliver global solutions for systems of nonlinear PDEs of the generality of those in (2.1), but it can also obtain a general, type independent, universal or blanket minimal *regularity* result for such solution, namely, it can prove that the solutions obtained can be assimilated with usual measurable functions, or even with *Hausdorff continuous* functions, and in fact, with smooth functions, under appropriate smoothness conditions on the respective nonlinear PDEs.

Clearly, as one of the consequences of solving nonlinear systems of PDEs of the generality of those in (2.1), the order completion method in [21,22,1-3,47-56,58,64-66] is the *only one* so far which fully manages to overcome the Lewy impossibility, [12].
And furthermore, it does so with a *large* nonlinear margin.

9. Order Completion Abolishes the Dichotomy "Linear Versus Nonlinear"

The *dichotomy linear versus nonlinear* relating to equations or operators in general is in its essence an issue of *algebra*, and more specifically, of *vector space* structures. In this way, it is present *both* in the functional analytic *and* algebraic methods - provided that they include vector spaces - for solving PDEs. And needless to say, dealing with the nonlinear case proves to be incomparably more difficult than it is with the linear one. Consequently, and unfortunately, the presence of this dichotomy is one of the *major disadvantages* of both the functional analytic and certain algebraic methods in solving PDEs, even if by now it is taken so much for granted that no wider ranging and more systematic attempt ever seems to be made to overcome it, let alone abolish what appears to be its inevitable presence.

On the other hand, order structures are of a more *basic* type than the algebraic ones.

Consequently, if instead of algebraic structures we consider order structures on the spaces of smooth functions on which the partial differential operators act naturally when considered in the classical context, see for instance (8.2), then these order structures - being more basic than algebra - *can no longer distinguish* between the linearity or nonlinearity of such partial differential operators.
In this way, the traditional dichotomy between the linear and nonlinear is simply set aside, and then the only problem left is whether indeed one can solve PDEs in the completion of such order structures.

Fortunately, as shown in [21,22,1-3,47-56,58,64-66], such a solution is possible and useful.

Here of course, one may think that it may help if the respective partial differential operators are monotonous. And then one may be concerned that all we managed to do was simply to get rid of the dichotomy between linear and nonlinear, so that instead, now we have to face the dichotomy monotonous versus arbitrary partial differential operators, with the latter being quite likely again far more difficult to deal with, than the former.
Such a particular approach in which the dichotomy monotonous versus non-monotonous prevails has recently been pursued in [5], for instance, with the consequent and not surprising considerable limitations on the types of PDEs to which it can be applied.

This however is clearly *not* the way with the method in [21,22,1-3,47-56,58,64-66]. Indeed, although in this method order structures and order completions are essentially used in solving PDEs, one nevertheless need *not* at all require a priori any sort of monotonicity property related to the equations solved. Certainly, the generality of nonlinear PDEs in (2.1), or of systems of such nonlinear PDEs, clearly illustrates the fact that the respective equations are *not* supposed to satisfy any a priori monotonicity conditions whatsoever.

What happens is very simple in fact, and perhaps surprisingly, is *similar* with the way the operators $T(x, D)$ and $\mathcal{T}$ in (8.3) acquire continuity type properties when the functional analytic method is used in the construction of such commutative diagrams. Indeed, with the functional analytic method, when one starts, say, with the natural mappings (8.2), the uniform topologies one considers on the classical domains and ranges of the operators in order to obtain commutative diagrams (8.3) are not arbitrary, but typically are related to the respective operators, see section 6 above, or [21, chap. 12].

The very same happens when order completion is used in [21,22,1-3,47-56,58,64-66] for the construction of commutative diagrams (8.3). More precisely, the order structures on the spaces of smooth functions on which the partial differential operators naturally act, see for instance (8.2), will typically be defined *depending* to a certain extent on these operators. In this way, the respective operators, no matter how arbitrary within the class of those in (2.1), will nevertheless *become* monotonous in the thus resulting partial orders, therefore, so much easier to deal with.

Such a procedure obviously cannot be imitated within algebra, since a nonlinear operator cannot in general define a vector space structure in which it would become linear.

Furthermore, the mentioned procedure used in [21,22,1-3,47-56,58,64-66] obviously goes far beyond the approach in [5], for instance, where one starts with given natural order relations on *both* the domains and ranges of ODEs or PDEs, and then severely restricts oneself only to those rather small classes of equations whose associated operators, or rather, inverse operators, are monotonous in the a priori given orders.

10. The Hidden Power of Methods Based on Partial Orders

As it happens, there is a rather widespread perception in mathematics that order structures are far too simple, and thus powerless, especially in analysis, therefore, they can deliver less than algebra, which on its turn, can deliver less than topology.

Accordingly, since the emergence of functional analytic methods in the solution of PDEs at the beginning of the 20th century, with the respective wealth of topologies on a variety of spaces of functions, the perception prevails that there simply cannot be any other more powerful methods in present day mathematics which could deal with such equations.
This perception has obviously further been strengthened by the fact that - lacking actually a sufficient ability to deal in a type independent manner with solving linear and nonlinear PDEs - the power of functional analytic methods got fragmented into dealing with a larger and larger variety more and more particular types of linear and nonlinear PDEs. And needless to say, such an increasing fragmentation has led to more and more powerful results, each of them with a less and less

wider applicability.

Since that process has by now been ongoing ever since the 1930s, it has entrenched the perception that there is not, and there cannot be any other way forward ...

In this regard the two examples of that perception mentioned in section 1 above are highly typical, being in fact promoted by what goes by the name of "leading" mathematicians ...

In view of such a perception it may therefore appear rather surprising, if not in fact hardly credible, to see results such as in [1-3,47-56,58,64-66] and the very first ones in [21,22], results obtained through order structures, and which results regarding PDEs could so far not be approached anywhere near by functional analytic methods.

Indeed, this method - based on order completion - and introduced in [21,22], does solve systems of PDEs of the nonlinear generality of those in (2.1), together with associated initial and/or boundary value problems, and furthermore, delivers for them global solutions which can be assimilated with usual measurable, or even Hausdorff continuous functions. And those solutions can be yet more smooth, under corresponding smoothness conditions on the respective PDEs, [52].

It is in this way that the order completion method is not only unprecedented, but it may also look rather strange in view of the mentioned perception in mathematics related to order structures.

Therefore, one should indeed address the apparent *secret of the power* of the order completion method in solving such large classes of systems of nonlinear PDEs, together with their associated initial and/or boundary value problems.

And one can indeed do so simply by questioning the mentioned perception ...

This can perhaps best be done by the presentation of certain classical - even if less well known - examples of remarkable mathematical results which illustrate the power of order structures in yielding what usually are called *deep theorems*.

In this regard a rather impressive, yet less well known fact is given by

the 1936 "Spectral Theorem" of Freudenthal, see [13, chap. 6].
Let us recall here in short some of its rather deep consequences.

The mentioned "Spectral Theorem" is a theorem about partially ordered structures, and it was proved by Freudenthal exclusively in terms of such structures. Yet, what is of special relevance is that by suitable particularizations, one can obtain from it the following three results which are in fields as diverse as Operator Theory, Measure Theory, and linear PDEs :

- the celebrated spectral representation theorem for normal operators in Hilbert spaces,

- the highly nontrivial Radon-Nikodym theorem in measure theory, and

- the Poisson formula for harmonic functions in an open circle.

11. Conclusions

The unprecedented power of the order completion method in solving very general systems of nonlinear PDEs and the associated initial and/or boundary value problems, and obtaining solutions given by usual functions, stems from two facts :

- Partial orders are more basic mathematical structures than algebra or topology. And being more basic than algebra, partial orders do *not* distinguish between linear and nonlinear equations, operators, and so on. Consequently, partial orders treat the linear and nonlinear cases in the same manner. Functional analytic method clearly cannot do the same.

- When using order completion for solving PDEs, one need *not* assume any monotonicity properties of the respective equations.

In the order completion method, the partial orders on the spaces of functions which are the domains of definition of the partial differential operators considered are defined in relation to these operators. This is similar to the way the topologies on such domains are defined, when functional analytic methods are used. Further details in this regard can be found in [21, chap. 12, 13] and [21,22,1-3,47-56,58,64-66].

12. Appendix : Trying to Ease ... the Pains of ... Paradigm Rigidity ... :-) :-) :-)

As noted at the beginning in the Brief Sketch of the Book, as well as in sections 0 and 1, it is a ... perfectly natural human behaviour ... to manifest ... paradigm rigidity ..., and in general, ... competence rigidity ...

Poor good old economics professor Joseph E Stiglitz of Columbia University, New York, and recipient of the 2001 Nobel Prize, has for evermore tried seemingly tirelessly to convince us that, inevitably, the future will simply FORCE us humans into an era of ... continual learning ...

God forbid !!!

Indeed, when in Genesis 3:23 we read about poor Adam that :

> "Therefore the LORD God sent him forth from the garden of Eden, to till the ground from whence he was taken"

it is clear that, ever since, we humans were sentenced to ... continual work ...
Yes, continual ... hard work with our muscles, with our bodies ...

But with our minds, as well ???
No way !!!
No, at those ancient times in the beginning, God did not get really so ... upset ... by Adam for having eaten from that forbidden apple ...

So that, why all of a sudden, and exactly nowadays, when at long long last modern science and the technology based on it appear so much to promise to us an ever more easy life, there comes some American

professor - even if with a Nobel Prize - and so even more than outrageously and totally out of the blue, DECLARES the inevitability of ... continual learning ... ???
Yes, of nothing less than continual learning which, at least as we humans happen to understand ourselves, is still a ... GREATER ... punishment, than ... continual hard work ... ???

But then, who can really know ?
Many, far too many prognoses of future turned out never to materialize ...
After all, as the saying would have it : the hardest thing to foretell is the ... future ...

And yet, is there any kind of guarantee that, in spite of his Nobel Prize, Prof. Stiglitz has - hopefully - got it all wrong, completely wrong, in fact ?

And in case he did not, the issue is NOT so much related to him ...
After all, he already got his own ... prize ..., and can simply go and retire ...
No, the issue, I am terribly afraid, is just about everybody else ...

Yes, just imagine, yes indeed, please, try, and imagine what an extraordinary ever ongoing EFFORT would such a ... continual learning ... place on everybody who would not be in a position somehow to be able to avoid it !!!
And certainly, that fast coming era would most definitely NOT mean merely ... learning more and more about the very same subject ... !!!
No, not at all !!!
What it would mean INSTEAD - and actually, we can already see it happening to some extent - is PERIODICALLY to learn about certain RADICALLY new, and again and again new subjects ... !!!

Yes indeed, so far, and ever since Adam was so unceremoniously shown the door out of the Garden of Eden, well, it is TRUE that we have had to ... continually keep working with our bodies ...
However, with our ... bodies ONLY ... !!!
And thus, as far as ... learning ... was concerned, well, we may have

to learn ONCE - while still young - one or two things, and then GET IT OVER WITH learning for the WHOLE the rest of our lives ... !!!

But now ???
Now, we are THREATENED with nothing short of ... continual learning ... !!! More precisely ... BOTH with good old continual working AND the suddenly introduced continual learning ...

Yes, nothing short of a true ... horror of horrors ...

And we do not even know what kind of ... damned apple ... did we happen to bite into lately, in order to deserve such a terrible punishment ...

Yes indeed, all of that so suddenly and newly emerged horror would - among many other things - be the END of our eternal human indulgence in such ... mental ... luxuries, like ... paradigm rigidity ..., and in general, ... competence rigidity ...

Good for Prof. Stiglitz to have already gotten his prize before that new and soon to come era may hit him ...
After all, who can know whether he himself would be able to stand up to the mentioned ... horrors ... he himself is prognosticating ...

But let us come back to the ... NOW ..., or if possible, and for the special sake of our dear American reader friends, even to the ... RIGHT NOW ... !

Well, ... right now ..., we seem to face - as earlier mentioned - two truly considerable TROUBLES :

> - the pains of ... paradigm rigidity ...
>
> and
>
> - the fact that analysts, and in particular, functional analysts, do quite naturally happen to be crowding rather the "worm" and "frog" end of the spectrum of mathematical thinking, vision and intuition, than that of the "bird" end ...

Now, regarding the second trouble above, it seems that one simply cannot do much : one quite likely has a very very strong, if not in fact, ultimately inborn propensity for being in one's nature significantly more near to one of the alternatives of "worm", "frog" or "bird", than to the other two ones ...

On the other hand - and most dearly hoping the Prof. Stiglitz is NOT asking of us humans the impossible - we may perhaps be indeed able to switch to ... continual learning ...
Or at least, many enough of us may hopefully be able to do so ...
And that of course can only mean to ... keep learning for evermore RADICALLY new and new things ...
Thus in particular, to get to ... survive ... the pains of paradigm rigidity, and in general, of competence rigidity ...
Yes, to survive those pains, get accustomed to them, and even thrive on them to some extent ...
Just as we managed for long long ages by now to do with the ... curse ... of ... continual hard body work ...

But then clearly, the issue of those ... pains of paradigm rigidity and competence rigidity ... is to be seen better, and hopefully also dealt with, in the future ...

Meanwhile, and as a sort of ... relaxation ..., we may even try to have some fun with the second above trouble, one which anyhow, seems to be intractable ...

In this regard, as mentioned in section 0, the item "Birds and Frogs" in the Notices of the AMS, 2009, February, Volume 56, Number 2, pages 212-223, by Dyson, is indeed worth reading. And as it is somewhat longer, it can deal with a certain appropriateness with a number of important mathematicians of the last several generations.

As for arguing for their respective classification as "birds" or "frogs" - as Dyson does not consider the category of "worms" as well - he is quite to the point, except for one obviously most blatant case, namely, of John von Neumann, whom Dyson declares to be a "frog", and definitely not a "bird".

An amusing related fact is that, Dyson, who is merely two decades younger than von Neumann, had a good opportunity to know von Neumann near enough. Yet he seems to miss the truly unique visionary value - both theoretical, as well as most practical - of certain contributions of von Neumann. And given the fact that Dyson has proved himself to be widely familiar with science, as well as an original thinker, one may wonder whether the sloppy treatment he gives for von Neumann may, perhaps, be no more than an attempt to settle some old grudges, grudges which would have been harder to settle earlier, while there were still many enough living mathematicians who had known von Neumann well enough ...
Of course, Dyson is ready to ... pay the price ... of qualifying himself only as a "frog", and not as a "bird" ...
But then, can that indeed compensate for placing von Neumann in the same "frog" category ?
And nowadays, with the passing of time, and with von Neumann no longer among us for nearly six decades, fewer and fewer mathematician, physicists or computer science people, are sufficiently aware in enough detail about the truly revolutionary contributions he made to various branches of mathematics, physics, and computational theory, as well as their effective and most important applications ...
Consequently, it is easier to place in the public domain a derogatory judgment about von Neumann, and expect it to stick ...

In view of the above, I considered it necessary to write a critical item in the Notices of the American Mathematical Society, which got published under the title "Separating Mathematicians",in Vol. 56, No. 6, June/July 2009, pp. 688-689.
As usual in such occasions, following immediately my item, an alleged reply of Dyson was also published.
Amusingly however, Dyson tried his best to avoid the issues raised in my mentioned criticism ...
And needless to say, I did not try to pursue further the debate with him ...
Instead, I placed a related paper entitled "John von Neumann and Self-Reference ...", on both

http://hal.archives-ouvertes.fr/hal-00749831 and
http://viXra.org/abs/1211.0036
which for convenience, is reproduced here :

John von Neumann and Self-Reference ...

Elemér E Rosinger
Department of Mathematics and Applied Mathematics
University of Pretoria
Pretoria
0002 South Africa
eerosinger@hotmail.com

> Dedicated to Marie-Louise Nykamp

Abstract

It is shown that the description as a "frog" of John von Neumann in a recent item by the Princeton celebrity physicist Freeman Dyson does among others miss completely on the immensely important revolution of the so called "von Neumann architecture" of our modern electronic digital computers.

Was indeed John von Neumann a mere "frog", as Freeman Dyson classifies him ?

Let us give an example of gross omission of self-reference in a recent publication by an assumed Princeton celebrity, Freeman Dyson (b. 1923).

Originating from England, Dyson started with mathematics, and then switched to physics.
His claim to fame comes from his contribution to quantum electrodynamics which he made back in 1949.
Since 1953, he has been at the Institute for Advanced Study, Princeton, New Jersey, USA.

In the February 2009, Vol. 56, No. 2, pp.212-223, of the Notices of the American Mathematical Society, he has the item "Birds and Frogs". The item is the written version of Dyson's "AMS Einstein Lecture" of October 2008, a lecture which in fact was canceled.

In it, more or less appropriately, he segregates mathematicians into two sharply different categories, namely, bird, and on the other hand, frogs.

Among the "birds" who are supposed to have a wider vision he mentions at the beginning Descartes, while as "frogs" - who are supposed to live in the mud below and see only the flowers that grow nearby - he starts with Francis Bacon (1561-1626).

And with some apparent modesty, Dyson classifies himself as a "frog"
...

What is amusing, however, is that he labels John von Neumann as a "frog", too...

And here, assuming naturally that Dyson is fully honest, one can only see this classification of von Neumann as an utter lack of even a mere elementary understanding by Dyson of the truly revolutionary and fundamental use of self-reference by von Neumann.

But before going into some detail, it is worth mentioning that, ever since the Paradox of the Liar in ancient Greece, Western civilization has had nothing short of a horror of self-reference. And that horror was further entrenched into our modern times when, in 1903, Bertrand Russell reformulated that ancient paradox in terms of Set Theory, thus further helping in creating an immense problem in the Foundations of Mathematics.

Consequently, it may simply happen that Dyson never came to give any thought to the issue of self-reference, considering that it had been settled for good, ever since ancient Greece...

Be it as it may related to Dyson, the fact is, and so it is ever since ancient times, that in the Old Testament - not a less important pillar of Western civilization, than ancient Greek art, science and philosophy, or the ancient Roman legal, political and military systems - there is no trace whatsoever of the least reservation regarding self-reference. And on the contrary, in Exodus 3:14, it is nothing less than the name of God, in the formulation "I am that I am".

The above, needless to say, should not be construed as placing any obligation upon Dyson. After all, modern mathematicians, physicists, or for that matter, other practitioners of hard sciences, do not usually excel in their deeper knowledge of the roots of Western civilization...

Not so with von Neumann, however.

Indeed, one of by far most important novelties in our times is the introduction of electronic digital computers. And nowadays, there is a near universal dominance in the construction of such computers of what is called the "von Neumann architecture".
This, briefly means the following.
Two inputs are introduced in every such computer, namely, the "program" and the "data". And the computer is supposed to process the "data" according to the "program", and then give as an output the "results".

Well, before the present day computers built according to the "von Neumann architecture", there have been some rather sophisticated electrical computers, among them the one built by the American Herman Hollerith (1860-1929). This computer did in 1890 process the whole American census in only one year, while in 1880, and prior to the Hollerith computer, the census took no less than eight years to be processed.
The massive success of the Hollerith computer is shown among other by the fact that in 1924, under the presidency of Thomas J Watson, the IBM, that is, International Business Machines Corporation was founded in order to produce and spread such computers.

And then, what was the truly revolutionary novelty, one of a massive practical advantage, which the "von Neumann architecture" introduced in the world of computers ?

Simple indeed :

All the earlier computers, including the Hollerith, operated only and only upon the given "data", and did so according to the given "program" which remained the same during the whole computation.

The essence of the "von Neumann architecture", on the other hand, is that the computer can operate both on the "data" and the "program" itself, before obtaining the "results". And the way the computer operates on the "program" is dependent on the "data".

It follows therefore that here we have a clear and rather simple example of self-reference : the "program" acts upon itself, and does so according to its own structure, as well as the information in the "data".

And this simple self-reference was perfectly sufficient in order to unleash all the miracles of modern computation...

At the same time, it seems nevertheless to escape completely the awareness of Dyson...

But the story does not stop here :

A few years later, von Neumann showed that one can construct finite cellular automata which can reproduce themselves. Thus they may be used in spreading civilization beyond Planet Earth.

Here, however, one should note from the beginning that at first sight - the issue is highly nontrivial. Indeed, a self-reproducing automaton must, among others, contain the "program" of its own self-reproduction. And then, this "program of self-reproduction" must on its turn contain a program of its own self-reproduction, that is, a "program of self-reproduction of the program of self-reproduction" ...

And so it comes that we are facing an infinite sequence of such "programs" ...

Well, von Neumann showed that a rather simple finite cellular automaton can avoid the need for such an infinite construction...

And again, Dyson happened to miss on that, too...

Last, and not least, one should note the following :

The so called "von Neumann architecture" makes our electronic digital computers not quite perfectly self-referential, since the way any given "program" acts upon itself depends not only on the respective "program", but also on the given "data".

On the other hand, the self-referentiality of self-reproducing automata is indeed a pure and perfect self-referentiality, since it has nothing else involved in it, except for itself.

Back to Dyson, however...

Well, having missed utterly on both self-referentialities above, not to mention on their immense importance, be it actual or potential, he manages to find one of the many lectures von Neumann gave, a lecture to which allegedly von Neumann went unprepared...

Yes indeed, Dyson seems to be a "frog" ...

And how much can a "frog" understand a "bird" ... ?

Anyhow, von Neumann, in a research career of about a mere quarter of century, from which his other engagements during WW II took a lot of time, managed to obtain fundamental contributions to Game Theory, Foundations of Set Theory, Quantum Mechanics, Operator Theory, among others...

Quite some "frog", one would say...

But then, Dyson's handicap is not only the fact that he is indeed a "frog", having done very little remarkable in physics, except for his 1949 breakthrough, but he is also a physicist...
And as such, he is not supposed to understand much enough about

mathematics, and thus, about mathematicians...

Yes, honesty seems not to be enough, not even when coming from a physicist...

But until he may reach next year the venerable age of ninety, he may hopefully have some time to ponder about such issues...

....

So much for now about the TWO above issues, namely :

> - the existence of "birds", "frogs" and "worms", an issue which, so far, no one seems to challenge, although it can easily be misjudged,

> - the most regrettable phenomenon of "paradigm rigidity" - and in general "competence rigidity" - which some, like Prof. Stiglitz for instance, started to challenge quite powerfully ...

....

As if possibly related to "paradigm rigidity" - and in general "competence rigidity" - it may be mentioned that the paper [58] which is mostly the same with the above sections 0 - 11, was about a decade ago submitted for publication to the Transactions of AMS, Bulletin of AMS, and Mathematical Intelligencer. And amusingly, it was not accepted by them, without however any comment regarding the mathematical content ...

But then, there is no need to ... overreaction ...
Indeed, the paper [19], which is also about a rather unified and general approach to PDEs was - thank God ! - accepted by the Mathematical Intelligencer ...

It seems, therefore, that perhaps there not a ... unified and general ... paradigm hostility ... manifesting itself ...

Or who knows, the paper [19] is still based exclusively on the use of functional analysis ...

And then, possibly, it did not really ring so many ... paradigm bells ... as they paper [58] apparently did ...

References

[1] Anguelov R : Dedekind order completion of C(X) by Hausdorff continuous functions. Quaestiones Mathematicae, 2004, Vol. 27, No. 2, 1533-169

[2] Anguelov R, Rosinger E E : Hausdorff continuous solutions of nonlinear PDEs through the order completion method. Quaestiones Mathematicae, Vol. 28, 2005, 1-15, arXiv:math.AP/0406517

[3] Anguelov R, Rosinger E E : Solving large classes of nonlinear systems of PDEs. Computers and Mathematics with Applications. Vol. 53, No. 3-4, Feb. 2007, 491-507, arXiv:math/0505674

[4] Arnold V I : Lectures on PDEs. Springer Universitext, 2004

[5] Carl S, Heikkilä S : Nonlinear Differential Equations in Ordered Spces. Chapman & Hall/CRC, Monographs and Surveys in Pure and Applied Mathematics, VOl. 111, Boca Raton, 2000

[6] Colombeau J-F : Elementary Introduction to New Generalized Functions. North-Holland Mathematics Studies, Vol. 113, Amsterdam, 1985

[7] Ehrenpreis L : Solutions of some problems of division I. Amer. J. Math., Vol. 76, 1954, 883-903

[8] Evans L C : Partial Differential Equations. AMS Graduate Studies in Mathematics, Vol. 19, 1998

[9] Grosser M, Kunzinger M, Oberguggenberger M, Steinbauer R : Geometric Theory of Generalized Functions with Applications to General Relativity. Kluwer, Dordrecht, 2002

[10] Kuhn T S : The Structure of Scientific Revolutions. Univ. Chicago, 1962

[11] Kuhn T S : The Structure of Scientific Revolutions : 50th Anniversary Edition. Univ. Chicago, 2012

[12] Lewy, H : An example of smooth linear partial differential equation without solutions. Ann. Math., vol. 66, no. 2, 1957, 155-158

[13] Luxemburg W A J, Zaanen A C : Riesz Spaces, Vol. I. North-Holland, Amsterdam, 1971

[14] MacNeille H M : Partially ordered sets. Trans. AMS, vol. 42, 1937, 416-460

[15] Malgrange B : Existence et approximation des solutions des equations aux derivees partielles et des equations de convolutions. Ann. Inst. Fourier Grenoble, Vol. 6, 1955-56, 271-355

[16] Mallios A, Rosinger E E [1] : Abstract differential geometry, differential algebras of generalized functions, and de Rham cohomology. Acta Appl. Math., vol. 55, no. 3, Feb. 1999, 231-250

[17] Mallios A, Rosinger E E [2] : Space-time foam dense singularities and de Rham cohomology. Acta Appl. Math., vol. 67, 2001, 59-89, arxiv:math/0406540

[18] Mallios A, Rosinger E E [3] : Dense singularities and de Rham cohomology. In (Eds. Strantzalos P, Fragoulopoulou M) Topological Algebras with Applications to Differential Geometry and Mathematical Physics. Proc. Fest-Colloq. in honour of Prof. Anastasios Mallios (16-18 September 1999), pp. 54-71, Dept. Math. Univ. Athens Publishers, Athens, Greece, 2002

[19] Neuberger J W : Prospects for a central theory of partial differential equations. The Mathematical Intelligencer, Vol. 27, No. 3, Summer 2005, 47-55

[20] Oberguggenberger M B : Multiplication of Distributions and Applications to PDEs. Pitman Research Notes in Mathematics, Vol. 259. Longman, Harlow, 1992

[21] Oberguggenberger M B, Rosinger E E : Solution of Continuous Nonlinear PDEs through Order Completion. North-Holland Mathematics Studies, Vol. 181. North-Holland, Amsterdam, 1994

[22] See review MR 95k:35002

[23] Oxtoby J C : Measure and Category. Springer, New York, 1971

[24] Rosinger E E [1] : Embedding of the $\mathcal{D}'$ distributions into pseudo-topological algebras. Stud. Cerc. Math., Vol. 18, No. 5, 1966, 687-729.

[25] Rosinger E E [2] : Pseudotopological spaces, the embedding of the $\mathcal{D}'$ distributions into algebras. Stud. Cerc. Math., Vol. 20, No. 4, 1968, 553-582.

[26] Rosinger E E [3] : Division of Distributions. Pacif.J. Math., Vol. 66, No. 1, 1976, 257-263

[27] Rosinger E E [4] : Nonsymmetric Dirac distributions in scattering theory. In Springer Lecture Notes in Mathematics, Vol. 564, 391-399, Springer, New York, 1976

[28] Rosinger E E [5] : Distributions and Nonlinear Partial Differential Equations. Springer Lectures Notes in Mathematics, Vol. 684, Springer, New York, 1978.

[29] Rosinger E E [6] : Nonlinear Partial Differential Equations, Sequential and Weak Solutions, North Holland Mathematics Studies, Vol. 44, Amsterdam, 1980.

[30] Rosinger E E [7] : Generalized Solutions of Nonlinear Partial Differential Equations. North Holland Mathematics Studies, Vol. 146, Amsterdam, 1987.

[31] Rosinger E E [8] : Nonlinear Partial Differential Equations, An Algebraic View of Generalized Solutions. North Holland Mathematics Studies, Vol. 164, Amsterdam, 1990.

[32] Rosinger E E [9] : Global version of the Cauchy-Kovalevskaia theorem for nonlinear PDEs. Acta Appl. Math., Vol. 21, 1990, 331–343.

[33] Rosinger E E [10] : Characterization for the solvability of nonlinear PDEs, Trans. AMS, Vol. 330, No. 1, March 1992, 203-225.

[34] Rosinger E E [11] : Nonprojectable Lie Symmetries of nonlinear PDEs and a negative solution to Hilbert's fifth problem. In (Eds. N.H. Ibragimov and F.M. Mahomed) Modern Group Analysis VI, Proceedings of the International Conference in the New South Africa, Developments and Applications, Johannesburg, January 1996, 21-30. New Age Inter. Publ., New Delhi, 1997

[35] Rosinger E E [12] : Parametric Lie Group Actions on Global Generalised Solutions of Nonlinear PDEs, Including a Solution to Hilbert's Fifth Problem. Kluwer Acad. Publishers, Amsterdam, Boston, 1998

[36] Rosinger E E [13] : Arbitrary Global Lie Group Actions on Generalized Solutions of Nonlinear PDEs and an Answer to Hilbert's Fifth Problem. In (Eds. Grosser M, Hörmann G, Kunzinger M, Oberguggenberger M B) Nonlinear Theory of Generalized Functions, 251-265, Research Notes in Mathematics, Chapman & Hall / CRC, London, New York, 1999

[37] Rosinger E E [14] : How to solve smooth nonlinear PDEs in algebras of generalized functions with dense singularities (invited paper) Applicable Analysis, vol. 78, 2001, 355-378, arXiv:math/0406594

[38] Rosinger E E [15] : Differential Algebras with Dense Singularities on Manifolds. Acta Applicandae Mathematicae. Vol. 95, No. 3, Feb. 2007, 233-256, arXiv:math/0606358

[39] Rosinger E E [16] : Space-Time Foam Differential Algebras of Generalized Functions and a Global Cauchy-Kovalevskaia Theorem. Acta Applicandae Mathematicae, DOI 10.1007/s10440-008-9326-z, Received: 5 February 2008 / Accepted: 23 September 2008, 05.10.2008, vol. 109, no. 2, pp. 439-462, arxiv:math/0607397

[40] Rosinger E E [17] : Survey on Singularities and Differential Algebras of Generalized Functions : A Basic Dichotomic Sheaf Theoretic Singularity Test. Lambert Academic Publishing, Saarbr´ucken, 2014, with preliminary version at hal-00510751, version 1, 23 August, 2010

[41] Rosinger E E, Rudolph M : Group invariance of global generalised solutions of nonlinear PDEs : A Dedekind order completion method. Lie Groups and their Applications, Vol. 1, No. 1, July-August 1994, 203-215

[42] Rosinger E E, Walus E Y [1] : Group invariance of generalized solutions obtained through the algebraic method. Nonlinearity, Vol. 7, 1994, 837-859

[43] Rosinger E E, Walus E Y [2] : Group invariance of global generalised solutions of nonlinear PDEs in nowhere dense algebras. Lie Groups and their Applications, Vol. 1, No. 1, July-August 1994, 216-225

[44] See reviews MR 92d:46098, Zbl. Math. 717 35001, MR 92d:46097, Bull. AMS vol.20, no.1, Jan 1989, 96-101, MR 89g:35001

[45] Schwartz L : Sur l'impossibilite de la multiplication des distributions. C.R. Acad. Sci. Paris, vol. 239, 1954, 847-848

[46] Treves F : Applications of distributions to PDE theory. Amer. Math. Monthly, March 1970, 241-248

[47] Van der Walt J H : Generalized solutions to nonlinear first order Cauchy problems. UPWT2007/15, Universit of Pretoria.

[48] Van der Walt J H : The uniform order convergence structure on ML(X). Qaestiones Mathematicae, 31 (2008), 55-77

[49] Van der Walt J H : The order completion method for systems of nonlinear PDEs: Pseudo- topological perspectives. Acta Applicanda Mathematicae, 103 (2008), 1-17

[50] Van der Walt J H : The order completion method for systems of nonlinear PDEs revisited. Acta Applicanda Mathematicae 106 (2009), 149-176

[51] Van der Walt J H : On the completion of uniform convergence spaces and an application to nonlinear PDEs. Qaestiones Mathematicae, 32 (2009), 371-395

[52] Van der Walt J H : Solutions of smooth nonlinear PDEs. UPWT2009/02, Abstract and Applied Analysis (2011), ID 658936, 37 pages electronic article. University of Pretoria

[53] Van der Walt J H : The order completion method for systems of nonlinear PDEs: Solutions of initial value problems, UPWT2009/03, University of Pretoria

[54] Van der Walt J H : Singularities in Space-times Foam Algebras. Applicable Analysis 90 (2011), 1763-1774.

[55] Van der Walt J H : The order completion method for system of nonlinear PDEs: Solutions of initial value problems. Abstract and Applied Analysis, (2013), Article ID 739462, 12 pages electronic article.

[56] D. Agbebaku D, Anguelov R, Van der Walt J H : Hausdorff continuous solutions of conservation laws. In MD Todorov (editor), Proceedings of the 4th International Conference on Application of Mathematics in Technical and Natural Sciences (AMiTaNS11), (1 16 June 2012, St Constantine and Helena, Bulgaria), American Institute of Physics AIP Conference Proceedings 1487, 2012, pp 151-158, ISBN: 978-0-7354-1099-2

Some of further nonlinear PDE related web site papers of E E Rosinger :

[57] How to solve smooth nonlinear PDEs in algebras of generalized functions with dense singularities, arXiv:math/0406594

[58] Can there be a general nonlinear PDE theory for the existence of solutions ? arXiv:math/0407026

[59] Genuine Lie semigroups and semi-symmetries of PDEs, arXiv:math/0604561

[60] New Symmetry Groups for Generalized Solutions of ODEs, arXiv:math/0605109

[61] Differential Algebras with Dense Singularities on Manifolds, arXiv:math/0606358

[62] Which are the Maximal Ideals ? arXiv:math/0607082

[63] Space-Time Foam Differential Algebras of Generalized Functions and a Global Cauchy-Kovalevskaia Theorem, arXiv:math/0607397

[64] Solving General Equations by Order Completion, arXiv:math/0608450

[65] Extending Mappings between Posets, arXiv:math/0609234,

[66] New Method for Solving Large Classes of Nonlinear Systems of PDEs, arXiv:math/0610279

[67] Junction Conditions, Resolution of Singularities and Nonlinear Equations of Physics, arXiv:math/0611445

[68] How to Define Global Lie Group Actions on Functions, arXiv:math/0611486

[69] Improvement on a Central Theory of PDEs, arXiv:math/0703515

[70] Local Functions : Algebras, Ideals, and Reduced Power Algebras, arXiv:0912.4049

[71] Inevitable Infinite Branching in the Multiplication of Singularities, arXiv:1002.0938

[72] Reflections on an Asymmetry on the Occasion of Arnold's Passing Away ..., vixra:1008.0011

[73] PDEs and Symmetry : an Open Problem. http://hal.archives-ouvertes.fr/hal-00540779, http://vixra.org/abs/1011.0069, posted in November 2010

[74] Why the Colombeau Algebras Cannot Formulate, Let Alone Prove the Global Cauchy-Kovalevskaia Theorem ? http://hal.archives-ouvertes.fr/hal-00590876, http://vixra.org/abs/1105.0007, posted in May 2011

[75] Why the Colombeau Algebras Cannot Handle Arbitrary Lie Groups ? http://hal.archives-ouvertes.fr/hal-00591354, http://vixra.org/abs/1105.0009, posted in May 2011

Part II : Six Papers on the Order Completion Method

1. A Brief Announcement about the Increased Blanket Regularity of Solutions, and the Definition of Their Property

Hausdorff Continuous Solutions of Arbitrary Continuous Nonlinear PDEs through the Order Completion Method,
arXiv:math/0405546

Elemér E Rosinger

Department of Mathematics
and Applied Mathematics
University of Pretoria
Pretoria
0002 South Africa
eerosinger@hotmail.com

Abstract

In 1994 we showed that very large classes of systems of nonlinear PDEs have solutions which can be assimilated with usual measurable functions on the Euclidean domains of definition of the respective equations. Recently, the *regularity* of such solutions has significantly been improved by showing that they can in fact be assimilated with Hausdorff continuous functions. The method of solution of PDEs is based on the Dedekind order completion of spaces of smooth functions which are defined on the domains of the given equations. In this way, the method does *not* use functional analytic approaches, or any of the customary distributions, hyper-functions, or other generalized functions.

Type independent existence and regularity results for large classes of systems of nonlinear PDEs

Ten years ago, in [4], the following significant *threefold* breakthrough was obtained with respect to solving large classes of nonlinear PDEs, see MR 95k:35002. Namely :

a) arbitrary nonlinear PDEs of the form

(1) $$F(x, U(x), \ldots, D^p U(x), \ldots) = f(x), \quad x \in \Omega$$

with F jointly continuous in all it arguments, f in a class of measurable functions, $\Omega \subseteq \mathbb{R}^n$ arbitrary open, $p \in \mathbb{N}^n$, with $|p| \leq m$, for $m \in \mathbb{N}$ arbitrary given, and the unknown function $U : \Omega \longrightarrow \mathbb{R}$, were proven to have

b) solutions U which can be assimilated with usual measurable functions on Ω, and

c) the solution method was based on the Dedekind order completion of suitable spaces of smooth functions on Ω.

In fact, the conditions at a) can further be relaxed by assuming that F may admit certain *discontinuities*, [???].

The method of order completion and the results on the existence and regularity of solutions can easily be extended to *systems* of nonlinear PDEs of the above form (1). Furthermore, initial and/or boundary value problems can be dealt with easily by this order completion method.

In this way, the solutions of the unprecedented large class of nonlinear PDEs in (1) can be obtained *without* the use of any sort of distribution, hyper-functions, generalized functions, or of methods of functional analysis. Moreover, one obtains a general, *blanket regularity*, given by the fact that the solutions constructed can be assimilated with usual measurable functions on the corresponding domains Ω in Euclidean spaces.

Recently, in collaboration with R. Anguelov, [1], a further significant improvement of the above mentioned 1994 results was obtained. Namely, this time we can further improve the *regularity* properties of the solutions by proving that they always belong to the significantly smaller class of Hausdorff continuous functions on the open domains Ω, see Appendix for a short account on Hausdorff continuous functions.

It should be noted that the results in [4] on existence of solutions do for the first time in the literature manage fully to overcome the celebrated 1957 Hans Lewy impossibility, see [5], and in fact do so with a large nonlinear margin.

Also, the existence results in [4], and thus their mentioned recent improvement with respect to the regularity of solutions, when solving large classes of nonlinear PDEs, *supersede* to a good extent the earlier similar ones obtained through the algebraic nonlinear theory of generalized functions introduced by the author in the 1960s, and developed since then alone or in collaborations, see 46F30 in the AMS Subject Classification at www.ams.org/index/msc/46Fxx.html, as well as [6], [7], [3, p. 7] and the literature cited there, or MR 89g:35001, MR 92d:46098, Zbl.Math.717 35001, Bull.AMS, Jan. 1989, Vol. 20, No. 1, 96-101.

To further facilitate the understanding of the above mentioned results, it may be useful to point to the following. In his latest 2004 edition of his Springer Universitext book "Lectures on PDEs", see [2], V I Arnold starts on page 1 with the statement :

"In contrast to ordinary differential equations, there is *no unified theory* of partial differential equations. Some equations have their own theories, while others have no theory at all. The reason for this complexity is a more complicated geometry ..." (italics added)

However, as the above mentioned results show, since [4], there is an *existence and regularity* of solutions theory for the large class of systems of nonlinear PDEs of the form in (1). Moreover, recently, the regularity result has been significantly improved by proving the existence of Hausdorff continuous solutions for such general nonlinear systems of PDEs.

Appendix. Hausdorff continuous functions

We shall deal with functions whose values can be usual or *extended* real numbers, that is, elements in

$$\overline{\mathbb{R}} = \mathbb{R} \cup \{-\infty, +\infty\}$$

Moreover, we shall allow the values of the functions to be not only numbers in $\overline{\mathbb{R}}$, but also *closed intervals* of such numbers, namely

$$[a, b] \subseteq \overline{\mathbb{R}}, \quad a, b \in \overline{\mathbb{R}}, \quad a \leq b$$

It turns out to be quite surprising how much more appropriate such a framework is when one deals with large classes of *non-smooth* functions in what is usually called Real Analysis.

Indeed, by considering such *interval valued* functions one obtains a systematic and effective way to study and deal with a large variety of non-smooth functions. Furthermore, one can gain important insights into the properties of such non-smooth functions, properties which in fact are not available in the usual approach.

It appears therefore that with the emergence in the second half of the 19-th century of a rigorous approach to Analysis, and specifically, with the Dirichlet definition of a function as having values given only and only by one single number, a certain undesired limitation was imposed in an unintended manner, especially what the study of non-smooth functions is concerned.

Somewhat later, towards the end of the 19-th century, Baire brought in the concepts of *lower* and *upper semi-continuous* functions, when dealing with non-smooth real valued functions. And in effect, he as-

sociated with each real valued function f, *two* other real, or extended real valued functions $I(f)$ and $S(f)$, with $I(f) \leq f \leq S(f)$, which proved to be particularly helpful.
However, following the prevailing mentality, each of these three functions were considered as being single valued.

As it turns out, however, by considering *interval valued* functions, such as for instance $F(f) = [I(f), S(f)]$, one can significantly improve on the understanding and handling of non-smooth functions.
The study of interval valued functions can, among others, show that the particular case of functions which have values given by one single number is appropriate for continuous functions only.

On the other hand, non-smooth functions are much better described by suitably associated interval valued functions.

Indeed, in the case of functions f which are *not* continuous, a much better description can be obtained by considering them given by a *pair* of usual point valued functions, namely $f = [\,\underline{f}, \overline{f}\,]$, thus leading to interval valued functions. And then, a natural class which replaces, and also extends, the usual point valued continuous functions is that of *Hausdorff-continuous* interval valued functions. The distinctive and *essential* feature of these Hausdorff-continuous functions $f = [\,\underline{f}, \overline{f}\,]$ is a condition of *minimality* with respect to the *gap* between $\underline{f}$ and $\overline{f}$, with the further requirement that $\underline{f}$ be lower semi-continuous, and $\overline{f}$ be upper semi-continuous.

In retrospect, it is surprising to see how near Baire came to such a treatment of non-smooth functions, what deep results he obtained, and its correspondent for lower semi-continuous functions, and yet followed the prevailing trend which considered functions as having to have point, and not interval values.

A good measure of the *naturalness* of interval valued functions can be seen in the results related to the Dedekind order completion of various spaces of continuous functions. And it is precisely such recently obtained results which allow for the mentioned significantly increased regularity properties of solutions of PDEs.

These Dedekind order completions prove to be subspaces of Hausdorff-continuous, thus interval valued functions. By the way, the space of Hausdorff-continuous functions itself is order complete.

Such results extend easily to functions defined on large classes of topological spaces.
A further indication of the natural role interval values play in the study of non-smooth functions can be found in the Differential and Integral Calculus being presently developed for functions with such values.

It will be useful to start by introducing a few notations. Let

(A.1) $\quad \overline{\mathbb{IR}} = \{ [\underline{a}, \overline{a}] \mid \underline{a}, \overline{a} \in \overline{\mathbb{R}} = \mathbb{R} \cup \{-\infty, +\infty\}, \underline{a} \leq \overline{a} \}$

be the set of all finite or infinite closed intervals.

The functions which we consider can be defined on arbitrary topological spaces Ω. For the purposes of the nonlinear PDEs studied in this book, however, it will be sufficient to assume that $\Omega \subseteq \mathbb{R}^n$ are arbitrary open subsets.
Let us now consider the set of interval valued functions

(A2) $\quad \mathbb{A}(\Omega) = \{ f : \Omega \longrightarrow \overline{\mathbb{IR}} \}$

By identifying the point $a \in \overline{\mathbb{R}}$ with the degenerate interval $[a, a] \in \overline{\mathbb{IR}}$, we consider $\overline{\mathbb{R}}$ as a subset of $\overline{\mathbb{IR}}$. In this way $\mathbb{A}(\Omega)$ will contain the set of functions with extended real values, namely

(A3) $\quad \mathcal{A}(\Omega) = \{ f : \Omega \longrightarrow \overline{\mathbb{R}} \} \subseteq \mathbb{A}(\Omega)$

We define a partial order $\leq$ on $\overline{\mathbb{IR}}$ by

(A4) $\quad [\underline{a}, \overline{a}] \leq [\underline{b}, \overline{b}] \quad \Longleftrightarrow \quad \underline{a} \leq \underline{b}, \overline{a} \leq \overline{b}$

Now on $\mathbb{A}(\Omega)$ we define the partial order induced by (A2.1.4) in the usual point-wise way, namely, for $f, g \in \mathbb{A}(\Omega)$, we have

(A5) $\quad f \leq g \quad \Longleftrightarrow \quad f(x) \leq g(x), \quad x \in \Omega$

Clearly, when restricted to $\mathcal{A}(\Omega)$, the above partial order on $\mathbb{A}(\Omega)$ reduces to the usual one among point valued functions.

Given an interval $a = [\underline{a}, \overline{a}] \in \overline{\mathbb{IR}}$, we denote

(A6) $\quad w(a) \;=\; \begin{cases} \overline{a} - \underline{a} & \text{if } \underline{a}, \overline{a} \text{ finite} \\[4pt] \infty & \text{if } \begin{array}{l} \overline{a} = \infty \text{ and } \underline{a} \text{ finite, or} \\ \underline{a} = -\infty \text{ and } \overline{a} \text{ finite, or} \\ \underline{a} = -\infty \text{ and } \overline{a} = \infty \end{array} \\[12pt] 0 & \text{if } \underline{a} = \overline{a} = \pm\infty \end{cases}$

which is called the *width* of the interval a. Also, we denote by

(A7) $\quad |a| = \max\{\, |\underline{a}|, |\overline{a}| \,\}$

the *modulus* of the interval $a = [\underline{a}, \overline{a}] \in \overline{\mathbb{IR}}$.

In this way

(A8) $\quad \mathcal{A}(\Omega) \;=\; \{\, f \in \mathbb{A}(\Omega) \mid w(f(x)) = 0, \ \ x \in \Omega \,\} \;\subseteq\; \mathbb{A}(\Omega)$

Let $f \in \mathbb{A}(\Omega)$. For every $x \in \Omega$, the value of f is an interval, namely

$f(x) \;=\; [\, \underline{f}(x), \overline{f}(x) \,], \quad \text{with } \underline{f}(x), \overline{f}(x) \in \mathbb{R}, \ \ \underline{f}(x) \leq \overline{f}(x)$

Hence, every function $f \in \mathbb{A}(\Omega)$ can be written in the form

(A9) $\quad f = [\, \underline{f}, \overline{f} \,], \quad \text{with } \underline{f}, \overline{f} \in \mathcal{A}(\Omega), \ \ \underline{f} \leq f \leq \overline{f}$

and

(A10) $\quad f \in \mathcal{A}(\Omega) \quad \Longleftrightarrow \quad \underline{f} = f = \overline{f}$

In the particular case of functions in $\mathcal{A}(\Omega)$, that is, with extended real, but point, and not non-degenerate interval values, a number of results in the sequel were obtained by Baire.

Most of the more general results concerning functions in $\mathbb{A}(\Omega)$, that is, with values finite or infinite closed intervals, have recently been developed by Anguelov.

For $x \in \Omega$, we denote by $\mathcal{V}_x$ the set of all neighbourhoods $V \subseteq \Omega$ of x. Let us consider the pair of mappings $I, S : \mathbb{A}(\Omega) \to \mathcal{A}(\Omega)$, called *lower* and *upper Baire operators*, respectively, where for every function $f \in \mathbb{A}(\Omega)$, we define

(A11) $\quad I(f)(x) = \sup_{V \in \mathcal{V}_x} \inf \{ z \in f(y) \mid y \in V \}$

(A12) $\quad S(f)(x) = \inf_{V \in \mathcal{V}_x} \sup \{ z \in f(y) \mid y \in V \}$

In Baire, these two operators were considered and studied in the particular case of functions $f \in \mathcal{A}(\Omega)$.

In view of the main interest here in this book in *interval valued* functions $f \in \mathbb{A}(\Omega)$, it is useful to consider as well the following third mapping, namely, $F : \mathbb{A}(\Omega) \to \mathbb{A}(\Omega)$, defined by

(A13) $\quad F(f)(x) = [\, I(f)(x),\ S(f)(x)\,], \quad f \in \mathbb{A}(\Omega),\ x \in \Omega,$

and called the *graph completion operator*.

The lower and upper Baire operators I and S, and consequently, the graph completion operator F, applied to any interval valued function $f = [\, \underline{f},\ \overline{f}\,] \in \mathbb{A}(\Omega)$ can now be conveniently represented in terms of the functions $\underline{f}$ and $\overline{f}$. Indeed, from (A2.1.11), (A2.1.12) it is easy to see that

(A14) $\quad I(f) = I(\underline{f}), \quad S(f) = S(\overline{f})$

Hence $F(f)$ can be written in the form

(A15) $\quad F(f) = [\, I(f),\, S(f)\,] = [\, I(\underline{f}),\, S(\overline{f})\,]$

Let us note that for every function $f \in \mathbb{A}(\Omega)$ we have the relations, see (A5), (A9)

(A16) $\quad I(f) = I(\underline{f}) \leq \underline{f} \leq f \leq \overline{f} \leq S(\overline{f}) = S(f)$

and, thus, the inclusions

(A17) $\quad f(x) \subseteq F(f)(x), \quad x \in \Omega$

Furthermore, the lower Baire operator $I : f \to I(f)$, the upper Baire operator $S : f \to S(f)$ and the graph completion operator $F : f \to F(f) = [\, I(f),\, S(f)\,]$ are all monotone with respect to the order $\leq$ in (A5) on $\mathbb{A}(\Omega)$, which means that for every two functions $f, g \in \mathbb{A}(\Omega)$ we have

(A18) $\quad f \leq g \implies I(f) \leq I(g),\ S(f) \leq S(g),\ F(f) \leq F(g)$

The operator F is also monotone with respect to inclusion, namely

(A19) $\quad f(x) \subseteq g(x),\ x \in \Omega \implies F(f)(x) \subseteq F(g)(x),\ x \in \Omega$

With an immediate extension of Baire, one can also show that all three operators are idempotent, that is, for every $f \in \mathbb{A}(\Omega)$, we have

(A20) $\quad I(I(f)) = I(f),\quad S(S(f)) = S(f),\quad F(F(f)) = F(f)$

Definition A1

A function $f \in \mathbb{A}(\Omega)$ is called *segment-continuous*, or in short, *s-continuous*, if and only if

(A21) $\quad F(f) = f$

$\square$

In view of (A17), it is obvious that condition (A21) is equivalent with

(A21*) $F(f)(x) \subseteq f(x), \quad x \in \Omega$

Furthermore, (A20), (A21) give

(A22) $F(f)$ is s-continuous for $f \in \mathcal{A}(\Omega)$

Example A1

Let us illustrate the concept of s-continuity in the simplest case of functions with one variable and one single discontinuity. Thus, with $\Omega = \mathbb{R}$, we take $f : \Omega \longrightarrow \overline{\mathbb{R}}$, or in other words, $f \in \mathcal{A}(\Omega)$, defined by

$$f(x) = \begin{vmatrix} a & \text{if } x < 0 \\ b & \text{if } x = 0 \\ c & \text{if } x > 0 \end{vmatrix}$$

where $a, b, c \in \overline{\mathbb{R}}$, $a \neq c$. Then f is *not* s-continuous.

Let us now take $f : \Omega \longrightarrow \overline{\mathbb{IR}}$, that is, $f \in \mathbb{A}(\Omega)$, defined by

$$f(x) = \begin{vmatrix} a & \text{if } x < 0 \\ [\,b,\ c\,] & \text{if } x = 0 \\ d & \text{if } x > 0 \end{vmatrix}$$

where $a, b, c, d \in \overline{\mathbb{R}}$, $a \leq d$ and $b \leq c$. Then f is s-continuous, if and only if $b \leq a$ and $d \leq c$. Similarly, if $a \geq d$ and $b \leq c$, then f is s-continuous, if and only if $b \leq d$ and $a \leq c$.

Consequently, returning to the first example above, it follows that f is s-continuous, if and only if $a = b = c$, that is, if and only if f is continuous, see (A2.1.32) below, for the general case of functions

$f \in \mathcal{A}(\Omega)$.

□

The *fundamental* concept is presented now in

Definition A2

A function $f \in \mathbb{A}(\Omega)$ is called *Hausdorff-continuous*, or in short, *H-continuous*, if and only if f is s-continuous, and in addition, for every s-continuous function $g \in \mathbb{A}(\Omega)$, we have satisfied the *minimality* condition on f :

(A23) $g(x) \subseteq f(x), \quad x \in \Omega \quad \Longrightarrow \quad g = f$

We shall denote by $\mathbb{H}(\Omega)$ the set of all Hausdorff-continuous interval valued functions on Ω.

Example A2

Let us again consider the second function in Example A1 above. Then f is H-continuous, if and only if $a = b$ and $c = d$

Let us give three further examples.

First, let us define $\alpha : \mathbb{R} \longrightarrow \overline{\mathbb{IR}}$ by

$$\alpha(x) = \begin{vmatrix} -1 & \text{if } x < 0 \\ [-1, 1] & \text{if } x = 0 \\ 1 & \text{if } x > 0 \end{vmatrix}$$

and then we can define $\beta : \mathbb{R}^2 \longrightarrow \overline{\mathbb{IR}}$ by

$$\beta(x,\ y) = \begin{vmatrix} \alpha(\sin(1/(x^2+y^2))) & \text{if } (x,\ y) \neq (0,\ 0) \\ [\,-1,\ 1\,] & \text{if } (x,\ y) = (0,\ 0) \end{vmatrix}$$

It is easy to see that both α and β are H-continuous.

The third example is a typical *shock wave* solution of the well known nonlinear PDE in Fluid Dynamics

$$U_t + UU_x = 0, \quad t \geq 0,\ x \in \mathbb{R}$$

which corresponds to the initial value problem

$$U(0,\ x) = \begin{vmatrix} 1 & \text{if } x \leq -1 \\ -x & \text{if } -1 \leq x \leq 0 \\ 0 & \text{if } x \geq 0 \end{vmatrix}$$

Namely, with $\Omega = [0, \infty) \times \mathbb{R}$, we have the solution $U : \Omega \longrightarrow \overline{\mathbb{IR}}$ given by

$$U(t,\ x) = \begin{vmatrix} 1 & \text{if } 0 \leq t < 1,\ x < t - 1 \\ x/(t-1) & \text{if } 0 \leq t < 1,\ t - 1 \leq x \leq 0 \\ 0 & \text{if } 0 \leq t < 1,\ x > 0 \\ 1 & \text{if } t \geq 1,\ x < (t-1)/2 \\ [\,-1,\ 1\,] & \text{if } t \geq 1,\ x = (t-1)/2 \\ 0 & \text{if } t \geq 1,\ x > (t-1)/2 \end{vmatrix}$$

Then U is H-continuous.

Remark A1

The *minimality* condition (A23) in the above definition of H-continuous functions proves to play a fundamental role.

□

As for the significance of the *regularity* property of being Hausdorff continuous, here we an important *similarity* between usual continuous, and on the other hand, Hausdorff-continuous functions, on the other. Namely, both of them are determined *uniquely* if they are known on a *dense* subset of their domains of definition.

This property comes in spite of the fact that Hausdorff-continuous functions can have discontinuities on sets of first Baire category, and such sets can have arbitrary large positive Lebesgue measure.

Indeed, we have

Theorem A1

Let $f = [\,\underline{f},\,\overline{f}\,]$, $g = [\,\underline{g},\,\overline{g}\,] \in \mathbb{A}(\Omega)$ be two H-continuous functions, and suppose given any dense subset $D \subseteq \Omega$. Then with the partial order in (A5), we have

a) $\underline{f}(x) \leq \underline{g}(x),\ x \in D \quad \Longrightarrow \quad f \leq g$ on Ω

b) $\overline{f}(x) \leq \overline{g}(x),\ x \in D \quad \Longrightarrow \quad f \leq g$ on Ω

c) $f(x) \leq g(x),\ x \in D \quad \Longrightarrow \quad f \leq g$ on Ω

Also

d) $f(x) = g(x),\ x \in D \quad \Longrightarrow \quad f = g$ on Ω

References

1. Anguelov R, Rosinger E E : Solution of nonlinear PDEs by Hausdorff Continuous Functions (to appear)

2. Arnold V I : Lectures on PDEs. Springer Universitext, 2004

3. Grosser M, et.al. : Geometric Theory of Generalized Functions with Applications to General Relativity. Mathematics and its Applications, Vol. 573, Kluwer, Dordrecht, 2001

4. Oberguggenberger M B, Rosinger E E : Solutions of Continuous Nonlinear PDEs through Order Completion. North-Holland Mathematics Studies, Vol. 181, (432 pages). Amsterdam, 1994

5. Lewy H : An example of a smooth linear partial differential equation without solution. Ann. Math., Vol. 66, No. 2, 1957, 155-158

6. Rosinger E E : Parametric Lie Group Actions on Global Generalized Solutions of Nonlinear PDEs, including a Solution to Hilbert's Fifth Problem, (234 pages). Kluwer, Dordrecht, 1998

7. Rosinger E E : How to solve smooth nonlinear PDEs in algebras of generalized functions with dense singularities (invited paper). Applicable Analysis, Vol. 78, 2001, 355-378

2. More Detailed Presentation of the Stronger Blanket Regularity Property of Solutions

Hausdorff Continuous Solutions of Nonlinear PDEs through the Order Completion Method, arXiv:math/0406517

Roumen Anguelov and Elemer E Rosinger
Department of Mathematics and Applied Mathematics
University of Pretoria

SOUTH AFRICA
anguelov@scientia.up.ac.za
rosinger@scientia.up.ac.za

Abstract

It was shown in [13] that very large classes of nonlinear PDEs have solutions which can be assimilated with usual measurable functions on the Euclidean domains of definition of the respective equations. In this paper the regularity of these solutions has significantly been improved by showing that they can in fact be assimilated with Hausdorff continuous functions. The method of solution of PDEs is based on the Dedekind order completion of spaces of smooth functions which are defined on the domains of the given equations.

1. Introduction

The following significant *threefold* breakthrough was obtained in [13] with respect to solving large classes of nonlinear PDEs, see MR 95k:35002. Namely:

a) arbitrary nonlinear PDEs of the form

$$T(x,D)u = f(x), \quad x \in \Omega, \tag{1}$$

where

$$T(x,D)u = g(x, u(x), \ldots, D_x^p u(x), \ldots), \quad p \in \mathbb{N}^n, \ |p| \leq m, \tag{2}$$

with g jointly continuous in all it arguments, f in a class of measurable functions, $\Omega \subseteq \mathbb{R}^n$ arbitrary open, $m \in \mathbb{N}$ arbitrary given, and the unknown function $u : \Omega \longrightarrow \mathbb{R}$, were proven to have

b) solutions u which can be assimilated with usual measurable functions on Ω, and

c) the solution method was based on the Dedekind order completion of suitable spaces of smooth functions on Ω.

In fact, the conditions at a) can further be relaxed by assuming that g may admit certain *discontinuities*, namely, that it is continuous only on $(\Omega\setminus\Sigma)\times\mathbf{R}^{m^*}$, where Σ is a closed, nowhere dense subset of Ω, while m^* is the number of arguments in g minus n. This relaxation on the continuity of g may be significant since such subsets of discontinuity Σ can have arbitrary large positive Lebesgue measure.

In this way, the solutions of the unprecedented large class of nonlinear PDEs in (1) can be obtained *without* the use of any sort of distributions, hyper-functions, generalized functions, or of methods of functional analysis. Moreover, one obtains a general, *blanket regularity*, given by the fact that the solutions constructed can be assimilated with usual measurable functions on the corresponding domains Ω in Euclidean spaces.

In this paper we discuss a further significant improvement of the above mentioned results with respect to the *regularity* properties of the solutions. Namely, this time we show that they can be assimilated with the significantly smaller class of Hausdorff continuous functions on the open domains Ω. This improvement follows, among others, from a recent breakthrough, see [1], which solves a long outstanding problem related to the Dedekind order completion of spaces $C(X)$ of real valued continuous functions on rather arbitrary topological spaces X. The Hausdorff continuous functions are not unlike the usual real valued continuous functions. For instance, they assume real values on a dense subset of the domain and are completely determined by the values on this subset. However, these functions may also assume interval values on a certain subset of the domain. Hence the concept of Hausdorff continuity is formulated within the realm of interval valued functions. We denote by $\mathbb{A}(\Omega)$ the set of all functions defined on an open set $\Omega\subset\mathbf{R}^n$ with values which are finite or infinite closed real intervals, that is,

$$\mathbb{A}(\Omega) = \{f : \Omega \to \mathbb{IR}\},$$

where $\mathbb{IR} = \{[\underline{a},\overline{a}] : \underline{a},\overline{a}\in\overline{\mathbb{R}} = \mathbb{R}\cup\{\pm\infty\}, \underline{a}\leq\overline{a}\}$. Given an interval $a = [\underline{a},\overline{a}]\in\mathbb{IR}$,

$$w(a) = \begin{cases} \overline{a}-\underline{a} & \text{if } \underline{a},\overline{a} \text{ finite},\\ +\infty & \text{if } \underline{a}<\overline{a}=+\infty \text{ or } \underline{a}=-\infty<\overline{a},\\ 0 & \text{if } \underline{a}=\overline{a}=\pm\infty, \end{cases}$$

is the width of a, while $|a| = \max\{|\underline{a}|,|\overline{a}|\}$ is the modulus of a. An extended real interval a is called proper if $w(a) > 0$ and degenerate or point if $w(a) = 0$. Identifying $a \in \overline{\mathbb{R}}$ with the degenerate interval $[a,a] \in \mathbb{IR}$, we consider $\overline{\mathbb{R}}$ as a subset of $\mathbb{IR}$. In this way $\mathbb{A}(\Omega)$ contains the set of extended real valued functions, namely,

$$\mathcal{A}(\Omega) = \{f : \Omega \to \overline{\mathbb{R}}\}.$$

A partial order which extends the total order on $\overline{\mathbb{R}}$ can be defined on $\mathbb{IR}$ in more than one way. However, it will prove useful to consider on $\mathbb{IR}$ the partial order $\leq$ defined by

$$[\underline{a}, \overline{a}] \leq [\underline{b}, \overline{b}] \iff \underline{a} \leq \underline{b},\ \overline{a} \leq \overline{b}. \tag{3}$$

The partial order induced on $\mathbb{A}(\Omega)$ by (3) in a point-wise way, i.e.,

$$f \leq g \iff f(x) \leq g(x),\ x \in \Omega, \tag{4}$$

is an extension of the usual point-wise order in the set of extended real valued functions $\mathcal{A}(\Omega)$.

The application of Hausdorff continuous functions to problems in Analysis, e.g. [1], and to nonlinear PDEs as in this paper, are based on the quite extraordinary fact that the set $\mathbb{H}(\Omega)$ of all Hausdorff continuous functions on Ω is order complete while some of its important subsets are Dedekind order complete. We can recall that the usual spaces of real valued functions considered in Analysis or Functional Analysis, e.g. spaces of continuous functions, Lebesgue spaces, Sobolev spaces, are with very few exceptions neither order complete nor Dedekind order complete.

The definition of the concept of Hausdorff continuity and related terminology are discussed in Section 2. The Baire operators and the graph completion operator which are instrumental for the definition and the properties of Hausdorff continuous functions are also discussed in that section. In order to improve the readability of the paper, a short account of some basic properties of the Hausdorff continuous functions is given in the Appendix.

The use of extended real intervals in the definition of the set $\mathbb{A}(\Omega)$ is partially motivated by the fact that the Baire operators involve infimums and supremums which might not exists in the realm of the usual (finite) real intervals. However, the main motivation with regard to the present exposition is the need to accommodate solutions of

PDEs which are discontinuous at certain points of the domain Ω and unbounded in the neighborhood of these points, e.g. the so called finite time blow-up. For this purpose it will prove sufficient to consider only the nearly finite functions.

Definition 1 *A function $f \in \mathbb{A}(\Omega)$ is called nearly finite if there exists an open and dense subset D of Ω such that*

$$|f(x)| < +\infty, \ x \in D$$

After a brief introduction to the order completion method in Section 3 we give the main result of the paper in Section 4, namely, that the solutions of the equation (1) through the order completion method can be assimilated with nearly finite Hausdorff continuous functions.

2. Baire Operators, Graph Completion Operator and Hausdorff Continuity

For every $x \in \Omega$, $B_\delta(x)$ denotes the open δ-neighborhood of x in Ω, that is,

$$B_\delta(x) = \{y \in \Omega : ||x - y|| < \delta\}.$$

Let D be a dense subset of Ω. The pair of mappings $I(D, \Omega, \cdot)$, $S(D, \Omega, \cdot) : \mathbb{A}(D) \to \mathcal{A}(\Omega)$ defined by

$$I(D, \Omega, f)(x) = \sup_{\delta > 0} \inf\{z \in f(y) : y \in B_\delta(x) \cap D\}, x \in \Omega, \quad (5)$$

$$S(D, \Omega, f)(x) = \inf_{\delta > 0} \sup\{z \in f(y) : y \in B_\delta(x) \cap D\}, x \in \Omega, \quad (6)$$

are called lower Baire and upper Baire operators, respectively. Clearly for every $f \in \mathbb{A}(D)$ we have

$$I(D, \Omega, f)(x) \leq f(x) \leq S(D, \Omega, f)(x), \ x \in \Omega.$$

Hence the mapping $F : \mathbb{A}(D) \to \mathbb{A}(\Omega)$, called a graph completion operator, where

$$F(D, \Omega, f)(x) = [I(D, \Omega, f)(x), S(D, \Omega, f)(x)], \ x \in \Omega, \ f \in \mathbb{A}(\Omega), \tag{7}$$

is well defined and we have the inclusion

$$f(x) \subseteq F(f)(x), \quad x \in \Omega. \tag{8}$$

The name of this operator is derived from the fact that considering the graphs of f and $F(D, \Omega, f)$ as subsets of the topological space $\Omega \times \overline{\mathbb{R}}$, the graph of $F(D, \Omega, f)$ is the minimal closed set which is a graph of interval function on Ω and contains the the graph of f. In the case when $D = \Omega$ the sets D and Ω will be usually omitted from the operators' argument lists, that is,

$$I(f) = I(\Omega, \Omega, f), \quad S(f) = S(\Omega, \Omega, f), \quad F(f) = F(\Omega, \Omega, f)$$

Let us note that, the graph completion operator is monotone about inclusion with respect to the functional argument, that is, if $f, g \in \mathbb{A}(D)$ where D is dense in Ω then

$$f(x) \subseteq g(x), \ x \in D \implies F(D, \Omega, f)(x) \subseteq F(D, \Omega, g)(x), \ x \in \Omega. \tag{9}$$

Furthermore, the graph completion operator is monotone about inclusion with respect to the set D in the sense that if D_1 and D_2 are dense subsets of Ω and $f \in \mathbb{A}(D_1 \cup D_2)$ then

$$D_1 \subseteq D_2 \implies F(D_1, \Omega, f)(x) \subseteq F(D_2, \Omega, f)(x), x \in \Omega. \tag{10}$$

This, in particular, means that for any dense subset D of Ω and $f \in \mathbb{A}(\Omega)$ we have

$$F(D, \Omega, f)(x) \subseteq F(f)(x), x \in \Omega. \tag{11}$$

Let $f \in \mathbb{A}(\Omega)$. For every $x \in \Omega$ the value of f is an interval $[\underline{f}(x), \overline{f}(x)]$. Hence, the function f can be written in the form $f = [\underline{f}, \overline{f}]$ where $\underline{f}, \overline{f} \in \mathcal{A}(X)$ and $\underline{f} \leq \overline{f}$. The lower and upper Baire operators as well as the graph completion operator of an interval valued function $f = [\underline{f}, \overline{f}] \in \mathcal{A}(\Omega)$ can be conveniently represented in terms of the functions $\underline{f}$ and $\overline{f}$:

$$I(D, \Omega, f) = I(D, \Omega, \underline{f}), \quad S(D, \Omega, f) = S(D, \Omega, \overline{f}),$$

$$F(D, \Omega, f) = [I(D, \Omega, \underline{f}), S(D, \Omega, \overline{f})].$$

Definition 2 *A function $f \in \mathbb{A}(\Omega)$ is called Hausdorff continuous, or H-continuous, if for every function $g \in \mathbb{A}(\Omega)$ which satisfies the inclusion $g(x) \subseteq f(x)$, $x \in \Omega$, we have $F(g)(x) = f(x)$, $x \in \Omega$.*

The concepts of Hausdorff continuity is strongly connected to the concepts of semi-continuity of real functions. We have the following characterization of the fixed points of the lower and the upper Baire operators, see [11]:

$$I(f) = f \iff f - \text{lower semi-continuous on } \Omega \qquad (12)$$
$$S(f) = f \iff f - \text{upper semi-continuous on } \Omega \qquad (13)$$

Hence an interval function $f = [\underline{f}, \overline{f}]$ is H-continuous if and only if the following three conditions hold

(i) $\underline{f}$ is lower semi-continuous

(ii) $\overline{f}$ is upper semi-continuous

(iii) the set $\{\phi \in \mathcal{A}(\Omega) : \underline{f} \leq \phi \leq \overline{f}\}$ does not contain lower or upper semi-continuous functions other than $\underline{f}$ and $\overline{f}$

The concept of H-continuity can be considered as a generalization of the concept of continuity of real functions in the sense that the only real (point valued) functions contained in $\mathbb{H}(\Omega)$ are the continuous functions, that is,

$$\left. \begin{array}{c} f \in \mathcal{A}(\Omega) \\ f \text{ is H-continuous} \end{array} \right\} \implies f \text{ is continuous} \qquad (14)$$

The H-continuous functions retain some essential properties of the usual real continuous functions as stated in Theorem 9 in the Appendix. Further links with the real continuous functions are presented in Theorems 10 and 11. We should also note that any Hausdorff continuous function f is "essentially" point valued in the sense that it assumes point values everywhere except on a set W_f which is of first Baire category, see Theorem 8. Through an application of the Baire category theorem this implies that the complement of W_f in Ω is a set of second Baire category. Hence

$$D_f = \Omega \setminus W_f = \{x \in \Omega : f(x) \in \mathbb{R}\} \text{ is dense in } \Omega. \qquad (15)$$

3. Order Completion Method for Nonlinear PDEs

The order completion method in solving general nonlinear systems of PDEs of the form (1) is based on certain very simple, even if less than usual, approximation properties, see [13]. To give an idea about the ways the order completion method works, we mention some of these approximations here.

The differential operator $T(x, D)$ on the left hand side of (1) has the following basic approximation property :

Lemma 3

$\forall \ x_0 \in \Omega, \ \epsilon > 0 \ :$

$\exists \ \delta > 0, \ P \ \text{polynomial in} \ x \in \mathbf{R}^n \ :$

$$\|x - x_0\| \leq \delta \implies f(x) - \epsilon \leq T(x, D)P(x) \leq f(x)$$

$\square$

Consequently, we obtain :

Proposition 4

$\forall \ \epsilon > 0 \ :$

$\exists \ \Gamma_\epsilon \subset \Omega \ \text{closed, nowhere dense in} \ \Omega, \ U_\epsilon \in C^\infty(\Omega) \ :$

$$f - \epsilon \leq T(x, D)P \leq f \quad \text{on} \quad \Omega \setminus \Gamma_\epsilon$$

Furthermore, one can also assume that the Lebesgue measure of Γ_ϵ is zero, namely

$$\text{mes} \ (\Gamma_\epsilon) = 0.$$

$\square$

In view of Proposition 4, the spaces of piecewise smooth functions given by

$$C_{nd}^l(\Omega) = \left\{ u \middle| \begin{array}{l} \exists\, \Gamma \subset \Omega \text{ closed, nowhere dense }: \\ *)\ u : \Omega \setminus \Gamma \to \mathbf{R} \\ **)\ u \in C^l(\Omega \setminus \Gamma) \end{array} \right\} \quad (16)$$

where $l \in \mathbf{N}$, are considered. It is easy to see that we have the inclusion

$$T(x, D)\, C_{nd}^m(\Omega) \subseteq C_{nd}^0(\Omega) \quad (17)$$

The general existence result obtained in [13] is represented through the following equation

$$T(x, D)^{\#}\, (C_{nd}^m(\Omega))_T^{\#} = (C_{nd}^0(\Omega))^{\#} \quad (18)$$

Here $(C_{nd}^m(\Omega))_T^{\#}$ and $(C_{nd}^0(\Omega))^{\#}$ are Dedekind order completions of $C_{nd}^m(\Omega)$ and $C_{nd}^0(\Omega)$, respectively, when these latter two spaces are considered with suitable partial orders. The respective partial order on $C_{nd}^m(\Omega)$ may depend on the nonlinear partial differential operator $T(x, D)$ in (17), while the partial order on $C_{nd}^0(\Omega)$ is the natural pointwise one at the points where the two functions compared are both continuous. The operator $T(x, D)^{\#}$ is a natural extension of the nonlinear partial differential operator $T(x, D)$ in (17) to the mentioned Dedekind order completions.

Equation (18) means that for every right hand term $f \in (C_{nd}^0(\Omega))^{\#}$ in (1), there exists a solution $u \in (C_{nd}^m(\Omega))_T^{\#}$, and as seen later, the set $(C_{nd}^0(\Omega))^{\#}$ contains many discontinuous functions beyond those piecewise discontinuous.

There is an obvious ambiguity with the piecewise smooth functions in $C_{nd}^m(\Omega)$. Indeed, given any such function u, the corresponding closed, nowhere dense set Γ cannot be defined uniquely. Therefore, it is convenient to factor out this ambiguity. For the space $C_{nd}^0(\Omega)$ which is the largest of these spaces of functions and also the range of $T(x, D)$ in (17) this is done by defining on it the equivalence relation $u \sim v$ for any two elements $u, v \in C_{nd}^0(\Omega)$, as given by

$$u \sim v \iff \left[\begin{array}{l} \exists\, \Gamma \subset \Omega \text{ closed, nowhere dense:} \\ (i)\ u, v \in C(\Omega \setminus \Gamma) \\ (ii)\ u = v \text{ on } \Omega \setminus \Gamma \end{array} \right]. \quad (19)$$

The mentioned ambiguity is eliminated by going to the quotient space

$$\mathcal{M}^0(\Omega) = C^0_{nd}(\Omega)/\sim \qquad (20)$$

The partial order on $C^0_{nd}(\Omega)$ induces a partial order on the quotient space $\mathcal{M}^0(\Omega)$, namely, for any two $\mathbf{u}, \mathbf{v} \in \mathcal{M}^0(\Omega)$ we have

$$\mathbf{u} \leq \mathbf{u} \iff \left[\begin{array}{l} \exists\, u \in \mathbf{u},\, v \in \mathbf{v},\, \Gamma \subset \Omega \text{ closed, nowhere dense:} \\ \text{i)}\ u, v \in C(\Omega \setminus \Gamma) \\ \text{ii)}\ u \leq v \text{ on } \Omega \setminus \Gamma \end{array}\right]. \qquad (21)$$

Using similar manipulations, this time also involving the operator $T(x, D)$, the ambiguity in the domain of $T(x, D)$ in (17) is factored out, thus producing the space $\mathcal{M}^0_T(\Omega)$ with partial order which may also depend on the operator T. For details on this procedure see [13]. The equation (18) is now replaced by

$$T(x, D)^\# \, (\mathcal{M}^m_T(\Omega))^\#_T = (\mathcal{M}^0(\Omega))^\# \qquad (22)$$

The basic regularity result in [13] is obtained by embedding $(\mathcal{M}^0(\Omega))^\#$ in the set of all measurable functions on Ω. Hence the solutions of (1) can be assimilated with measurable functions.

In the next section we will show that $\mathcal{M}^0(\Omega)$ can be embedded in the set $\mathbb{H}(\Omega)$ of all Hausdorff continuous functions on Ω. Since the set $\mathbb{H}(\Omega)$ is order complete it also contains the Dedekind order completion of $\mathcal{M}^0(\Omega)$. More precisely, we obtain that $\mathcal{M}^0(\Omega)$ is order isomorphic to $\mathbb{H}_{nf}(\Omega)$. Hence the solutions of (1) can be assimilated with nearly finite Hausdorff continuous functions.

4. Further on the Order Completion Method for Nonlinear PDEs

Let $u \in C_{nd}(\Omega)$. According to (16), there exists a closed, nowhere dense set $\Gamma \subset \Omega$ such that $u \in C(\Omega \setminus \Gamma)$. Since $\Omega \setminus \Gamma$ is open and dense in Ω, we can define

$$F_0(u) = F(\Omega \setminus \Gamma, \Omega, u) \qquad (23)$$

The closed, nowhere dense set Γ used in (23), is not unique. However, we can show that the value of $F(\Omega \setminus \Gamma, \Omega, u)$ does not depend on the set Γ in the sense that for every closed, nowhere dense set Γ such that $u \in C(\Omega \setminus \Gamma)$ the value of $F(\Omega \setminus \Gamma, \Omega, u)$ remains the same.

Let Γ_1 and Γ_2 be closed, nowhere dense sets such that $u \in C(\Omega \setminus \Gamma_1)$ and $u \in C(\Omega \setminus \Gamma_2)$. Then the set $\Gamma_1 \cup \Gamma_2$ is also closed and nowhere dense. According to Theorem 11 in the Appendix the functions $F(\Omega \setminus \Gamma_1, \Omega, u)$, $F(\Omega \setminus \Gamma_2, \Omega, u)$ and $F(\Omega \setminus (\Gamma_1 \cup \Gamma_2), \Omega, u)$ are all H-continuous and for every $x \in \Omega \setminus (\Gamma_1 \cup \Gamma_2)$ we have

$$F(\Omega \setminus \Gamma_1, \Omega, u)(x) = F(\Omega \setminus \Gamma_2, \Omega, u)(x) = F(\Omega \setminus (\Gamma_1 \cup \Gamma_2), \Omega, u)(x) = u(x).$$

Since $\Omega \setminus (\Gamma_1 \cup \Gamma_2)$ is dense in Ω Theorem 9 implies that

$$F(\Omega \setminus \Gamma_1, \Omega, u) = F(\Omega \setminus \Gamma_2, \Omega, u) = F(\Omega \setminus (\Gamma_1 \cup \Gamma_2), \Omega, u).$$

Therefore, the mapping

$$F_0 : C_{nd}(\Omega) \longmapsto \mathbb{A}(\Omega)$$

is unambiguously defined through (23). In analogy with (7), we call F_0 a graph completion mapping on $C_{nd}(\Omega)$. As mentioned above already it follows from Theorem 11 that for every $u \in C_{nd}(\Omega)$ we have

$$F_0(u) \in \mathbb{H}(\Omega).$$

Furthermore, if $u \in C(\Omega \setminus \Gamma)$ we have

$$F_0(u)(x) = u(x), \ x \in \Omega \setminus \Gamma. \tag{24}$$

The above identity shows that the values of the function $F_0(u)$ are finite on the open and dense set $\Omega \setminus \Gamma$. Hence, $F_0(u)$ is nearly finite, see Definition 1. Thus, we have

$$F_0 : C_{nd}(\Omega) \longmapsto \mathbb{H}_{nf}(\Omega) \tag{25}$$

The following theorem shows that the images of two functions in $C_{nd}(\Omega)$ under the mapping F_0 are the same if and only if these functions are equivalent with respect to the relation (19).

Theorem 5 *Let $u, v \in C_{nd}(\Omega)$. Then*

$$F_0(u) = F_0(v) \iff u \sim v$$

Proof. <u>Implication to the left.</u> Let Γ be closed, nowhere dense subset of Ω associated with u and v in terms of (19), that is,

$$u, v \in C(\Omega \setminus \Gamma),$$
$$u(x) = v(x), \ x \in \Omega \setminus \Gamma.$$

The required equality follow from (23) where the set Γ is the one considered above. Indeed, we have

$$F_0(u) = F(\Omega \setminus \Gamma, \Omega, u) = F(\Omega \setminus \Gamma, \Omega, v) = F_0(v)$$

<u>Implication to the right.</u> Let us denote by Γ_1 and Γ_2 the closed, nowhere dense sets associated with the functions u and v, respectively, in terms of (16), that is, Γ_1 and Γ_2 are such that $u \in C(\Omega \setminus \Gamma_1)$ and $v \in C(\Omega \setminus \Gamma_2)$. Assume that

$$F_0(u) = F_0(v). \qquad (26)$$

The functions u and v are both continuous on the set $\Omega \setminus (\Gamma_1 \cup \Gamma_2)$. Therefore, from the property (24) and the assumption (26) it follows that

$$u(x) = F_0(u)(x) = F_0(v)(x) = v(x), \ x \in \Omega \setminus (\Gamma_1 \cup \Gamma_2).$$

Since the set $\Gamma_1 \cup \Gamma_2$ is closed and nowhere dense in Ω, the above identity implies that $u \sim v$, see (19). ∎

In view of (25) and Theorem 5 now we can define a mapping

$$\mathbf{F}_0 : \mathcal{M}^0(\Omega) \longmapsto \mathbb{H}_{nf}(\Omega)$$

in the following way. Let $\mathbf{u} \in \mathcal{M}^0(\Omega)$ and let $\phi \in \mathbb{H}_{nf}(\Omega)$. Then

$$\mathbf{F}_0(\mathbf{u}) = \phi \iff \exists u \in \mathbf{u} : F_0(u) = \phi. \qquad (27)$$

It is easy to see that the definition of $\mathbf{F}_0(\mathbf{u})$ does not depend on the particular representative u of the equivalence class $\mathbf{u}$. Indeed, if $u, h \in \mathbf{u}$ then $u \sim h$. Thus, $F_0(u) = F_0(h)$, see Theorem 5. Therefore the statement (27) can be reformulated as

$$\mathbf{F}_0(\mathbf{u}) = \phi \iff \forall u \in \mathbf{u} : F_0(u) = \phi. \qquad (28)$$

Theorem 6 *The mapping* $\mathbf{F}_0 : \mathcal{M}^0(\Omega) \longmapsto \mathbb{H}_{nf}(\Omega)$ *defined by (27) is an order isomorphic embedding with respect to the order relation (21) in $\mathcal{M}^0(\Omega)$ and the order relation (4) in $\mathbb{H}_{nf}(\Omega)$, that is, for any $\mathbf{u},\mathbf{v} \in \mathcal{M}^0(\Omega)$ we have*

$$\mathbf{u} \leq \mathbf{v} \Longleftrightarrow \mathbf{F}_0(\mathbf{u}) \leq \mathbf{F}_0(\mathbf{v}).$$

Proof. Let $\mathbf{u},\mathbf{v} \in \mathcal{M}^0(\Omega)$ and $\mathbf{u} \leq \mathbf{v}$. According to (21) there exists a closed, nowhere dense set Γ in Ω and functions $u \in \mathbf{u}$, $v \in \mathbf{v}$ such that $u, v \in C(\Omega \setminus \Gamma)$ and $u(x) \leq v(x)$ for all $x \in \Omega \setminus \Gamma$. Using the same set Γ in the evaluation of $F_0(u)$ and $F_0(v)$ according to (23) as well as the monotonicity of the graph completion operator, see Theorem 12, we have

$$F_0(u) = F(\Omega \setminus \Gamma, u) \leq F(\Omega \setminus \Gamma, v) = F_0(v)$$

Let us assume now that $\mathbf{u}, \mathbf{v} \in \mathcal{M}^0(\Omega)$ and $\mathbf{F}_0(\mathbf{u}) \leq \mathbf{F}_0(\mathbf{v})$. From the representation (27) of $\mathbf{F}_0$ it follows that there exist $u \in \mathbf{u}$ and $v \in \mathbf{v}$ such that $F_0(u) = \mathbf{F}_0(\mathbf{u})$ and $F_0(v) = \mathbf{F}_0(\mathbf{v})$. Obviously we have

$$F_0(u) \leq F_0(v). \tag{29}$$

Since $u, v \in C_{nd}(\Omega)$ there exist closed, nowhere dense sets Γ_1 and Γ_2 such that $u \in C_{nd}(\Omega \setminus \Gamma_1)$ and $v \in C_{nd}(\Omega \setminus \Gamma_2)$. The set $\Gamma = \Gamma_1 \cup \Gamma_2$ is also closed, nowhere dense. Both functions u and v are continuous on $\Omega \setminus \Gamma$. Therefore $F_0(u)(x) = u(x)$ and $F_0(v)(x) = v(x)$ for all $x \in \Omega \setminus \Gamma$, see (24). Hence, inequality (29) implies

$$u(x) \leq v(x), \; x \in \Omega \setminus \Gamma,$$

which means that $\mathbf{u} \leq \mathbf{v}$, see (21). ∎

Theorem 7 *Let $h \in \mathbb{H}_{nf}(\Omega)$. There exists a subset $\mathcal{G}$ of $\mathcal{M}^0(\Omega)$ such that $h = \sup \mathbf{F}_0(\mathcal{G})$, where $\mathbf{F}_0(\mathcal{G})$ is the range of $\mathcal{G}$ under $\mathbf{F}_0$, that is, $\mathbf{F}_0(\mathcal{G}) = \{\mathbf{F}_0(\mathbf{u}) : \mathbf{u} \in \mathcal{G}\}$.*

Proof. The set $\Gamma_{nf}(h) = \{x \in \Omega : \infty \in h(x) \text{ or } -\infty \in h(x)\}$ is closed, nowhere dense, see Definition 1, and $h \in \mathbb{H}_{ft}(\Omega \setminus \Gamma_{nf}(h))$. Then according to Theorem 14 the function h can be represented on the set $\Omega \setminus \Gamma_{nf}(h)$ as

$$h(x) = (\sup \mathcal{F})(x), \; x \in \Omega \setminus \Gamma_{nf}(h), \tag{30}$$

where
$$\mathcal{F} = \{v \in C(\Omega \setminus \Gamma_{nf}(h)) : v(x) \leq h(x) \, , \, x \in \Omega \setminus \Gamma_{nf}(h)\}.$$

The set $\mathcal{F}$ is a subset of $C_{nd}(\Omega)$ because $\Gamma_{nf}(h)$ is closed and nowhere dense. We will show that
$$h = \sup \mathbf{F}_0(\mathcal{G})$$
where
$$\mathcal{G} = \{\mathbf{v} \in \mathcal{M}^0(\Omega) : \exists v \in \mathcal{F} : v \in \mathbf{v}\}.$$

Indeed, since all functions in $\mathcal{F}$ are continuous on $\Omega \setminus \Gamma_{nf}(h)$, for every $\mathbf{v} \in \mathcal{G}$ and $v \in \mathbf{v}$ we have, see Theorem 10,
$$v(x) = F_0(v)(x) = \mathbf{F}_0(\mathbf{v})(x) \, , \, x \in \Omega \setminus \Gamma_{nf}(h). \qquad (31)$$

Hence
$$\mathbf{F}_0(\mathbf{v})(x) = v(x) \leq h(x) \, , \, x \in \Omega \setminus \Gamma_{nf}(h)\}.$$

Using that both $\mathbf{F}_0(\mathbf{v})$ and h are H-continuous on Ω we obtain from Theorem 9 that
$$\mathbf{F}_0(\mathbf{v})(x) \leq h(x) \, , \, x \in \Omega \, , \, \mathbf{v} \in \mathcal{G}.$$

Therefore, h is an upper bound of $\mathbf{F}_0(\mathcal{G})$. As a bounded subset of $\mathbb{H}_{nf}(\Omega)$ the set $\mathbf{F}_0(\mathcal{G})$ has a supremum in $\mathbb{H}_{nf}(\Omega)$, see Theorem 13. Let $g = \sup \mathbf{F}_0(\mathcal{G})$. Clearly
$$g \leq h. \qquad (32)$$

Furthermore, from (31) it follows that for every $v \in \mathcal{F}$ and the respective class $\mathbf{v} \in \mathcal{G}$ containing v we have
$$v(x) = \mathbf{F}_0(\mathbf{v})(x) \leq g(x) \, , \, x \in \Omega \setminus \Gamma_{nf}(h).$$

Hence, g is an upper bound of $\mathcal{F}$ on the set $\Omega \setminus \Gamma_{nf}(h)$ while h is the supremum of $\mathcal{F}$ on $\Omega \setminus \Gamma_{nf}(h)$. Therefore,
$$h(x) \leq g(x), \, x \in \Omega \setminus \Gamma_{nf}(h).$$

Using the H-continuity of g and h, from Theorem 9 we obtain that
$$h(x) \leq g(x), \, x \in \Omega.$$

This together with (32) shows that $h = g = \sup \mathbf{F}_0(\mathcal{G})$ which completes the proof. ∎

The Theorem 7 shows that $\mathbb{H}_{nf}(\Omega)$ is the smallest Dedekind order complete subset of $\mathbb{H}(\Omega)$ which contains the image of $\mathcal{M}^0(\Omega)$ under the order isomorphic embedding $\mathbf{F}_0$. Hence it is order isomorphic to the Dedekind order completion $\mathcal{M}^0(\Omega)^{\#}$ of $\mathcal{M}^0(\Omega)$. The mapping discussed in this section are illustrated on the following diagram, $\mathbf{F}_0^{\#}$ denoting the order isomorphism from $\mathcal{M}^0(\Omega)^{\#}$ to $\mathbb{H}(\Omega)$.

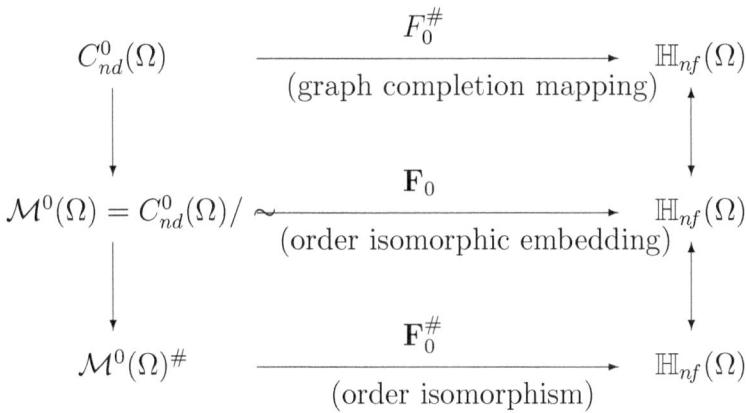

The set of solutions $\mathcal{M}_T^m(\Omega))_T^{\#}$ is mapped onto the set $\mathbb{H}_{nf}(\Omega)$ of all nearly finite Hausdorff continuous functions through the composition of the mappings $T(x, D)^{\#}$ and $\mathbf{F}_0^{\#}$. Considering that both mappings are order isomorphisms, the set of solutions $\mathcal{M}_T^m(\Omega))_T^{\#}$ is order isomorphic with the set $\mathbb{H}_{nf}(\Omega)$. Hence, the solutions of (1) through the order completion method can be assimilated with nearly finite Hausdorff continuous functions.

5. Conclusions

The paper deals with the regularity of the solutions of nonlinear PDEs obtained through the order completion method. We show that these solutions can be assimilated with Hausdorff continuous function, thus significantly improving the results in [13] with respect to the regularity properties of the solutions. The applications of the class of Hausdorff continuous functions discussed here as well as in other recent publications, [1], [2], [9], show that this class may play an important role in what is typically called Real Analysis. In particular, one may note that one of the main engines behind the development of the various spaces in Real and Abstract Analysis are the partial differential equations with the need to assimilate the various types of "weak" solutions. Since the solutions of very large classes of nonlinear partial differential equations can be assimilated with nearly finite Hausdorff continuous functions, the set of these functions might be a viable alternative to some of the presently used functional spaces (e.g. $L^p(\Omega)$, Sobolev spaces) with the advantage of being both more regular and universal.

6. Appendix

The concept of Hausdorff continuous interval valued functions was developed first within the theory of Hausdorff approximations of real functions, see [14]. The name is derived from the fact that for a Hausdorff continuous function $f = [\underline{f}, \overline{f}]$ the Hausdorff distance between the graphs of $\underline{f}$ and $\overline{f}$ is zero. Since the Hausdorff continuous functions are in general interval valued they are also studied as a part of the Interval Analysis, see [3], [2].

The minimality condition associated with the Hausdorff continuity, see Definition 2, requires that the graph of a Hausdorff continuous function is as 'thin' as possible, that is, the function assumes proper interval values only when it is necessary to ensure that the graph of this interval function is a closed subset of $\Omega \times \mathbb{R}$. As a result the set where a Hausdorff continuous function assumes proper interval values is small. The next theorem shows that this set is meager or a set of first Baire category, that is, a countable union of closed and nowhere dense sets.

Theorem 8 *The set $W_f = \{x \in \Omega : w(f(x)) > 0\}$ of all points where $f \in \mathbb{A}(\Omega)$ assumes proper interval values is a set of first Baire category.*

It may appear at first that the minimality condition in Definition 2 applies at each individual point x of Ω, thus, not involving neighborhoods. However, the graph completion operator F does appear in this condition. And this operator according to (7) and therefore (5) and (6) does certainly refer to neighborhoods of points in Ω, a situation typical, among others, for the concept of continuity. Hence the following property of the continuous functions is preserved.

Theorem 9 *Let f, g be H-continuous on Ω and let D be a dense subset of Ω. Then*

a) $f(x) \le g(x), \ x \in D \implies f(x) \le g(x), \ x \in \Omega$,
b) $f(x) = g(x), \ x \in D \implies f(x) = g(x), \ x \in \Omega$.

The following two theorems represent essential links with the usual point valued continuous functions.

Theorem 10 *Let $f = [\underline{f}, \overline{f}]$ be an H-continuous function on Ω.*
a) If $\underline{f}$ or $\overline{f}$ is continuous at a point $a \in \Omega$ then $\underline{f}(a) = \overline{f}(a)$.
b) If $\underline{f}(a) = \overline{f}(a)$ for some $a \in \Omega$ then both $\underline{f}$ and $\overline{f}$ are continuous at a.

Theorem 11 *Let D be a dense subset of Ω. If $f \in C(D)$ then*

$$F(D, \Omega, f) \in \mathbb{H}(\Omega),$$
$$F(D, \Omega, f)(x) = f(x), \ x \in D.$$

A partial order which extends the total order on $\mathbb{R}$ can be defined on $\mathbb{IR}$ in more than one way. Historically, several partial orders are associated with the set $\mathbb{IR}$, namely,
(i) the inclusion relation $[\underline{a}, \overline{a}] \subseteq [\underline{b}, \overline{b}] \iff \underline{b} \le \underline{a} \le \overline{a} \le \overline{b}$
(ii) the "strong" partial order $[\underline{a}, \overline{a}] \preceq [\underline{b}, \overline{b}] \iff \overline{a} \le \underline{b}$
(iii) the partial order defined by (3).
The use of the inclusion relation on the set $\mathbb{IR}$ is motivated by the applications of interval analysis to generating enclosures of solution sets. However, the role of partial orders extending the total order on $\mathbb{R}$ has also been recognized in computing, see [13]. Both orders (ii) and (iii) are extensions of the order on $\mathbb{R}$. The use of the order (ii) is based on the view point that inequality between intervals should imply inequality between their interiors. This approach is rather limiting since the order (ii) does not retain some essential properties of the order on $\mathbb{R}$. For instance, a proper interval A and the interval $A + \varepsilon$

are not comparable with respect to the order (ii) when the positive real number ε is small enough. The partial order (iii) is introduced and studied by Markov, see [12], [11]. The results reported in [1] and in the present paper indicate that indeed the partial order (4) induced pointwise by (3) is an appropriate partial order to be associated with the Hausdorff continuous interval valued functions.

The monotonicity with respect to the relation inclusion was discussed in Section 2, see (9) and (10). The following theorem states the monotonicity of the Baire operators and the graph completion operator with respect to the order (4) induces in a pontwise way by the order (3).

Theorem 12 *The lower Baire operator, the upper Baire operator and the graph completion operator are all monotone increasing with respect to the order (4) on the respective domains and ranges, that is, if D is a dense subset of Ω, for every two functions $f, g \in \mathbb{A}(D)$ we have*

$$f(x) \leq g(x), \ x \in D \implies \begin{cases} I(D, \Omega, f)(x) \leq I(D, \Omega, g)(x), \ x \in \Omega \\ S(D, \Omega, f)(x) \leq S(D, \Omega, g)(x), \ x \in \Omega \\ F(D, \Omega, f)(x) \leq F(D, \Omega, g)(x), \ x \in \Omega \end{cases}$$

Important property of the set $\mathbb{H}(\Omega)$ is that it is order complete. As we noted, the order completeness or the Dedekind order completeness is not a typical property for the spaces of functions considered in Real Analysis. In this way, the class of H-continuous functions and its subclasses mentioned in the next theorem can provide solutions to open problems or improve earlier results related to order.

Theorem 13 .

(i) The set $\mathbb{H}(\Omega)$ of all H-continuous functions is order complete.

(ii) The set $\mathbb{H}_{bd}(\Omega)$ of all bounded H-continuous functions, that is,

$$\mathbb{H}_{bd}(\Omega) = \{f \in \mathbb{H}(\Omega) : \exists M \in \mathbb{R} : |f(x)| \leq M, \ x \in \Omega\}$$

is Dedekind order complete

(iii) The set $\mathbb{H}_{ft}(\Omega)$ of all finite H-continuous functions, that is,

$$\mathbb{H}_{ft}(\Omega) = \{f \in \mathbb{H}(\Omega) : |f(x)| < +\infty, \ x \in \Omega\}$$

is Dedekind order complete

(iv) The set $\mathbb{H}_{nf}(\Omega)$ of all nearly finite H-continuous functions, that is,

$$\mathbb{H}_{nf}(\Omega) = \{f \in \mathbb{H}(\Omega) : \exists D- \text{ dense subset of } \Omega : |f(x)| < +\infty,\ x \in D\}$$

is Dedekind order complete.

The resent paper [1] gives the Dedekind order completion of the space $C(X)$ of all continuous real functions on a topological space X in terms of Hausdorff continuous functions, thus improving significantly an earlier result by Dilworth, see [14]. The main result in [1] is stated below for the case when $X = \Omega$.

Theorem 14 *The set $\mathbb{H}_{ft}(\Omega)$ is a Dedekind order completion of the set $C(\Omega)$. Moreover, for every $h \in \mathbb{H}_{ft}(\Omega)$ we have*

$$h = sup\{f \in C(\Omega) : f \leq h\}$$

The proof of the theorems in this apendix can be found in [1] and [2].

References

[1] R Anguelov, Dedekind Order Completion of C(X) by Hausdorff Continuous Functions, Quaestiones Mathematicae, to appear.

[2] R Anguelov, An Introduction to the Spaces of Interval Functions, Technical Report UPWT2004/4, University of Pretoria.

[3] R. Anguelov and S. Markov, Extended segment analysis, *Freiburger Intervall - Berichte* 10 (1981), 1 - 63.

[4] R. Anguelov, E.E. Rosinger, Solution of Nonlinear PDEs by Hausdorff Continuous Functions (to appear).

[5] R. Baire, Lecons sur les Fonctions Discontinues, Collection Borel, Paris, 1905.

[6] M Bardi, I Capuzzo-Dolcetta, *Optimal control and viscosity solutions of Hamilton-Jacobi-Bellman equations*, Birkhäuser, Boston, Basel, Berlin, 1997.

[7] G. Birkhoff, The Role of Order in Computing, in *Reliability in Computing* (ed. R. Moore) (Academic Press, 1988), 357–378.

[8] R. P. Dilworth, The normal completion of the lattice of continuous functions, *Trans. Amer. Math. Soc.* **68** (1950), 427–438.

[9] W.A.J. Luxemburg, A.C. Zaanen, Riesz Spaces I, North Holland, Amsterdam, 1971.

[10] S. Markov, A nonstandard subtraction of intervals, *Serdica* **3** (1977), 359–370.

[11] S. Markov, Calculus for interval functions of a real variable, *Computing* **22** (1979), 325–337.

[12] S. Markov, Extended interval arithmetic involving infinite intervals, *Mathematica Balkanica* **6** (1992), 269–304.

[13] M.B. Oberguggenberger, E.E. Rosinger, Solution on Continuous Nonlinear PDEs through Order Completion, North-Holland, Amsterdam, London, New York, Tokyo, 1994.

[14] B. Sendov, *Hausdorff Approximations* (Kluwer Academic, Boston, 1990).

3. Certain Details on Solving Large Classes of Nonlinear Systems of PDEs

Solving Large Classes of Nonlinear Systems of PDEs, arXiv:0505674

Roumen Anguelov & Elemér E Rosinger
Department of Mathematics
and Applied Mathematics
University of Pretoria
Pretoria

0002 South Africa
roumen.anguelov@up.ac.za
eerosinger@hotmail.com

Abstract

It is shown that large classes of nonlinear systems of PDEs, with possibly associated initial and/or boundary value problems, can be solved by the method of order completion. The solutions obtained can be assimilated with Hausdorff continuous functions. The usual Navier-Stokes equations, as well as their various modifications aiming at a realistic modelling, are included as particular cases. The same holds for the critically important constitutive relations in various branches of Continuum Mechanics. The solution method does not involve functional analysis, nor various Sobolev or other spaces of distributions or generalized functions. The general and type independent *existence* and *regularity* results regarding solutions presented here have recently been introduced in the literature.

> "... provided also if need be that the notion of a solution shall be suitably extended ..."
>
> cited from Hilbert's 20th Problem

1. Main ideas of the order completion solution method

The solution method is divided in two parts. The proof of the *existence* of solutions follows the method of order completion introduced and first developed in Oberguggenberger & Rosinger. The proof of the *regularity* of solutions is a consequence of recent results obtained in Anguelov [1], regarding the structure of the Dedekind order completion of spaces of continuous functions $\mathcal{C}(X)$, where X is a rather arbitrary topological space.

For simplicity of presentation, we shall consider single nonlinear PDEs. The extension to systems of such nonlinear PDEs and associated initial and/or boundary value problems can - rather surprisingly - be done easily, as seen in Oberguggenberger & Rosinger, this is being one of the major advantages of the order completion method. Let us therefore consider nonlinear PDEs of the general form

(1.1) $\quad F(x, U(x), \ldots, D_x^p U(x), \ldots) = f(x), \quad x \in \Omega \subseteq \mathbb{R}^n$

with $p \in \mathbb{N}^n$, $|p| \leq m$, where the domains Ω can be any open, not necessarily bounded subsets of $\mathbb{R}^n$, while the orders $m \in \mathbb{N}$ of the PDEs are arbitrary given, and the unknown functions, that is, the solutions one looks for are $U : \Omega \longrightarrow \mathbb{R}$.

The *unprecedented generality* of these nonlinear PDEs comes, above all, from the class of functions F which define the left hand terms, and which are only assumed to be *jointly continuous* in all of their arguments. The right hand terms f are also required to be *continuous*.

As seen, however, both F and f can have certain *discontinuities* as well, Oberguggenberger & Rosinger.

Regardless of the above generality of the nonlinear systems of PDEs considered, one can find for them solutions U defined on the *whole* of the respective domains Ω. These solutions U have the *blanket, type independent*, or *universal regularity* property that they can be assimilated with *Hausdorff continuous functions*.

It follows in this way that, when solving systems of nonlinear PDEs of the generality of those in (1.1), one can *dispense with* the various customary spaces of distributions, hyperfunctions, generalized functions, Sobolev spaces, and so on. Instead one can stay within the realms of *usual functions*, more precisely, of *interval valued* functions, see the Appendix for a short presentation of essentials on Hausdorff continuous functions. Also, when proving the *existence* and the mentioned type of *regularity* of such solutions one can dispense with methods of Functional Analysis. However, functional analytic methods can possibly be used in order to obtain further regularity or other desirable properties of such solutions.

Let us now associate with each nonlinear PDE in (1.1) the corresponding nonlinear partial differential operator defined by the left hand side, namely

(1.2) $\quad T(x, D)U(x) = F(x, U(x), \ldots, D_x^p U(x), \ldots), \quad x \in \Omega$

Two facts about the nonlinear PDEs in (1.1) and the corresponding nonlinear partial differential operators $T(x, D)$ in (1.2) are important and immediate :

- The operators $T(x, D)$ can *naturally* be seen as acting in the *classical* context, namely

(1.3) $\quad T(x, D) : \mathcal{C}^m(\Omega) \ni U \longmapsto T(x, D)U \in \mathcal{C}^0(\Omega)$

while, unfortunately on the other hand :

- The mappings in this natural classical context (1.3) are typically *not* surjective even in the case of linear $T(x, D)$, and they are even less so in the general nonlinear case of (1.1), (1.2).

In other words, linear or nonlinear PDEs in (1.1) typically *cannot* be expected to have *classical* solutions $U \in \mathcal{C}^m(\Omega)$, for arbitrary continuous right hand terms $f \in \mathcal{C}^0(\Omega)$, as illustrated by a variety of well known examples, some of them rather simple ones, see Oberguggenberger & Rosinger [chap. 6].

Furthermore, it can often happen that nonclassical solutions do have a major applicative interest, thus they have to be sought out *beyond* the confines of the classical framework in (1.3).

This is, therefore, how we are led to the *necessity* to consider *generalized solutions* U for PDEs like those in (1.1), that is, solutions $U \notin \mathcal{C}^m(\Omega)$, which therefore are no longer classical. This means that the natural classical mappings (1.3) must in certain suitable ways be *extended* to commutative diagrams

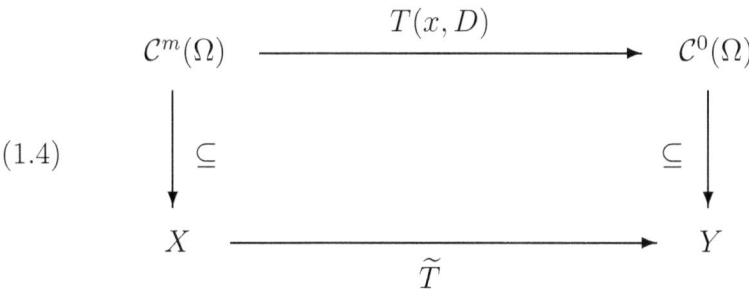

(1.4)

with the generalized solutions now being found as

(1.5) $\quad U \in X \setminus \mathcal{C}^m(\Omega)$

instead of the classical ones $U \in \mathcal{C}^m(\Omega)$ which may easily fail to exist. A further important point is that one expects to reestablish certain kind of *surjectivity* type properties typically missing in (1.3), at least such as for instance

(1.6) $\quad \mathcal{C}^0(\Omega) \subseteq \tilde{T}(X)$

As it turns out, when constructing extensions of (1.3) given by commutative diagrams (1.4), we shall be interested in the following somewhat larger spaces of piecewise smooth functions. For any integer $0 \leq l \leq \infty$, we define

(1.7) $\quad \mathcal{C}_{nd}^l(\Omega) = \left\{ u : \Omega \to \mathbb{R} \;\middle|\; \begin{array}{l} \exists\, \Gamma \subset \Omega \text{ closed, nowhere dense :} \\ u \in \mathcal{C}^l(\Omega \setminus \Gamma) \end{array} \right\}$

and as an immediate strengthening of (1.3), we obviously obtain

(1.8) $\quad T(x,D)\, \mathcal{C}_{nd}^m(\Omega) \subseteq \mathcal{C}_{nd}^0(\Omega)$

The solution of the nonlinear PDEs in (1.1) through the order completion method will come from the construction of specific instances of the *commutative diagrams* (1.4), given by, see (2.18), (2.27)

(1.9)
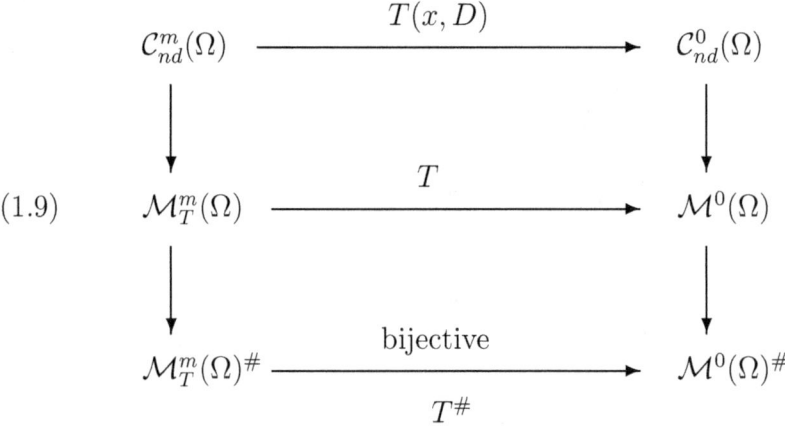

where, as elaborated later, the operation ( )$^\#$ means the *Dedekind order completion*, according to MacNeille, of the respective spaces, as well as the extension to such completions of the respective mappings, see Oberguggenberger & Rosinger [Appendix]. It follows that in terms of (1.4), we have

$X = \mathcal{M}_T^m(\Omega)^\#, \quad Y = \mathcal{M}^0(\Omega)^\#, \quad \tilde{T} = T^\#$

thus we shall obtain for the nonlinear PDEs in (1.1) generalized solutions

(1.10) $\quad U \in \mathcal{M}_T^m(\Omega)^\#$

Furthermore, instead of the *surjectivity* condition (1.6), we shall at least have the following stronger one

(1.11) $\quad \mathcal{C}_{nd}^0(\Omega) \subseteq T^\#(\mathcal{M}_T^m(\Omega)^\#)$

So far about the main ideas related to the *existence* of solutions of general nonlinear PDEs of the form (1.1).

As for the *regularity* of such solutions, we recall that, as shown in Oberguggenberger & Rosinger, one has the inclusions

(1.12) $\quad \mathcal{M}^0(\Omega)^\# \subseteq Mes(\Omega)$

where $Mes(\Omega)$ denotes the set of Lebesgue measurable functions on Ω. In this way, in view of (1.9), (1.10), one can assimilate the generalized solutions U of the nonlinear PDEs in (1.1) with usual measurable functions in $Mes(\Omega)$.

Recently, however, based on results in Anguelov [1], it was shown that instead of (1.12), one has the much stronger property

(1.13) $\quad \mathcal{M}^0(\Omega)^\# \subseteq \mathbb{H}(\Omega)$

where $\mathbb{H}(\Omega)$ denotes the set of Hausdorff continuous functions on Ω. Consequently, now one can significantly improve on the earlier regularity result, as one can assimilate the generalized solutions U of the nonlinear PDEs in (1.1) with usual functions in $\mathbb{H}(\Omega)$.

2. The construction of diagram (1.9)

Since we solve PDEs through order completion, let us see how near we can come to satisfying the equality in (1.1), when using inequalities. For that purpose, it is useful to consider for each $x \in \Omega$ the following set of real numbers

(2.1) $\quad \mathbb{R}_x = \{\, F(x, \xi_0, \ldots, \xi_p, \ldots) \mid \xi_p \in \mathbb{R},\ \text{for}\ p \in \mathbb{N}^n,\ |p| \leq m\,\}$

Clearly, for $x \in \Omega$ fixed, $\mathbb{R}_x$ is the range in $\mathbb{R}$ of $F(x, \ldots)$, and since F is jointly continuous in all its arguments, it follows that $\mathbb{R}_x$ is a nonvoid interval which is bounded, half bounded, or is the whole of $\mathbb{R}$. This latter case, which can happen often with nonlinear PDEs in (1.1), will be easier to deal with, see (2.3) next. Clearly, in the case of non-degenerate linear PDEs in (1.1), this latter case always happens. Given now $x \in \Omega$, it is obvious that a *necessary* condition for the existence of a classical smooth solution $U \in \mathcal{C}^m$ of (1.1) in a neighbourhood of x is the condition

(2.2) $\quad f(x) \in \mathbb{R}_x$

Consequently, for the time being, we shall make the assumption that the right hand term functions f in the nonlinear PDEs in (1.1) satisfy the somewhat stronger version of condition (2.2) given by

(2.3) $\quad f(x) \in \text{int } \mathbb{R}_x, \quad \text{for } x \in \Omega$

Clearly, whenever we have

(2.4) $\quad \mathbb{R}_x = \mathbb{R}, \quad \text{for } x \in \Omega$

then (2.3) is satisfied. And as mentioned, this is the case with all nontrivial linear PDEs, as well as with most of the nonlinear PDEs of practical interest.

And now the basic and rather simple *local approximation* result on how near we can satisfy the equality in (1.1), when using inequalities.

Proposition 2.1.

Given $f \in \mathcal{C}^0(\Omega)$, then

(2.5) $\quad \begin{array}{l} \forall \ x_0 \in \Omega, \ \epsilon > 0 \ : \\ \exists \ \delta > 0, \ P \text{ polynomial in } x \in \mathbb{R}^n \ : \\ \forall \ x \in \Omega, \ \|x - x_0\| \leq \delta \ : \\ f(x) - \epsilon \leq T(x, D)P(x) \leq f(x) \end{array}$

Proof

Given $x_0 \in \Omega$, then for $\epsilon > 0$ small enough, condition (2.3) yields $\xi_p \in \mathbb{R}$, with $p \in \mathbb{N}^n$, $|p| \leq m$, such that

(2.6) $\quad F(x_0, \xi_0, \ldots, \xi_p, \ldots) = f(x_0) - \epsilon/2$

Let us take P a polynomial in $x \in \mathbb{R}^n$, which satisfies the conditions

$D_x^p P(x_0) = \xi_p, \quad p \in \mathbb{N}^n, \ |p| \leq m$

In this case from (2.6) we clearly obtain the relation

(2.7) $\quad T(x_0, D)P(x_0) = f(x_0) - \epsilon/2$

and since both $T(x, D)P(x)$ and $f(x)$ are continuous in $x \in \Omega$, the local inequality property (2.5) follows easily from (2.7).

The *global approximation* version of the inequality property in (2.5) is given in

Proposition 2.2.

If $f \in \mathcal{C}^0(\Omega)$, then

(2.8) $\quad \begin{array}{l} \forall \ \epsilon > 0 \ : \\ \exists \ \Gamma_\epsilon \subset \Omega \text{ closed, nowhere dense, } U_\epsilon \in \mathcal{C}^m(\Omega \setminus \Gamma_\epsilon) \ : \\ f - \epsilon \leq T(x, D)U_\epsilon \leq f \text{ on } \Omega \setminus \Gamma_\epsilon \end{array}$

Proof

Let us take a covering of Ω of the form

(2.9) $\quad \Omega = \bigcup_{\nu \in \mathbb{N}} K_\nu$

where K_ν are compact n-dimensional intervals in $\mathbb{R}^n$, namely, $K_\nu = [a_\nu, b_\nu]$, with $a_\nu = (a_{\nu,1}, \ldots, a_{\nu,n})$, $b_\nu = (b_{\nu,1}, \ldots, b_{\nu,n}) \in \mathbb{R}^n$. We also assume, see Forster, that the covering (2.9) is locally finite, that is

(2.10) $\quad \forall \ x \in \Omega \ : \\ \quad \exists \ V_x \subseteq \Omega \ \text{neighbourhood of } x \ : \\ \quad \{\nu \in \mathbb{N} \mid K_\nu \cap V_x \ne \phi\} \ \text{is a finite set of indices}$

and furthermore

(2.11) the interiors of K_ν, with $\nu \in \mathbb{N}$, are pairwise disjoint

Let us now take $\epsilon > 0$ arbitrary but fixed. Further, we take $\nu \in \mathbb{N}$. We shall apply Proposition 2.1 to each $x_0 \in K_\nu$. Then we obtain $\delta_{x_0} > 0$ and a polynomial P_{x_0} such that

$$f(x) - \epsilon \le T(x,D)P_{x_0}(x) \le f(x), \quad x \in \Omega, \quad \|x - x_0\| \le \delta_{x_0}$$

But K_ν is compact, therefore

(2.12) $\quad \exists \ \delta > 0 \ : \\ \quad \forall \ x_0 \in K_\nu \ : \\ \quad \exists \ P_{x_0} \ \text{polynomial in } x \in \mathbb{R}^n \ : \\ \quad \forall \ x \in \Omega, \ \|x - x_0\| \le \delta_{x_0} \ : \\ \quad f(x) - \epsilon \le T(x,D)P_{x_0}(x) \le f(x)$

Now we shall subdivide K_ν, which was assumed to be a compact n-dimensional interval, into n-dimensional subintervals $I_1, \ldots, I_\mu$, so that the diameter of each of them is less or equal δ.

Let us denote by J a generic such n-dimensional subinterval in any of the K_ν, when $\nu \in \mathbb{N}$. If $a_J \in J$ is the center of any such n-dimensional subinterval then (2.12) gives for $x \in \text{int } J$

$$f(x) - \epsilon \le T(x,D)P_{a_J}(x) \le f(x)$$

Let us now take

(2.13) $\quad \Gamma_\epsilon = \Omega \setminus \bigcup_J \text{int } J$

that is, with the union ranging over all such n-dimensional subintervals J. In view of ??? If we define $U_\epsilon \in C^m(\Omega \setminus \Gamma_\epsilon)$ by

$U_\epsilon = P_{a_J} \ \text{on } \Omega \cap \text{int } J$

then the proof is completed.

Remark 2.1

1) It is easy to see that the inequalities in (2.5) and (2.8) can be replaced with the following ones, respectively

(2.14) $\quad f(x) \le T(x,D)P(x) \le f(x) + \epsilon$

(2.15) $\quad f \leq T(x,D)U_\epsilon \leq f+\epsilon$

as their proofs follow after the corresponding obvious minor changes in the proofs of the above two propositions. And these four inequalities are *sharper* than would respectively be the inequalities

$$f(x)-\epsilon \leq T(x,D)P(x) \leq f(x)+\epsilon, \quad f-\epsilon \leq T(x,D)U_\epsilon \leq f+\epsilon$$

As we shall see not much later, we do need the sharper inequalities. Indeed, the order completion method which we shall employ is based on MacNeille's construction, therefore, it uses *Dedekind cuts*. And such cuts do need the above sharper inequalities.

2) In Proposition 2.2., as well as in its version corresponding to the above inequality (2.15), we have in addition the property

(2.16) $\quad mes\,(\Gamma_\epsilon) = 0$

where *mes* denotes the usual Lebesgue measure. Indeed, according to (2.10), (2.11) and (2.13), Γ_ϵ is a countable union of rectangular grids, each generated by a finite number of hyperplanes.

Here it should be noted that the presence of the *closed, nowhere dense* singularity sets Γ_ϵ in the *global* inequalities (2.8) and (2.15) will prove not to be a hindrance. And in fact, it will lead to the classes of piecewise smooth functions in (1.7) which prove to be convenient.

The presence of such closed, nowhere dense *singularity* sets is rather deep rooted, as it is connected with such facts as the *flabbiness* of related sheaves of functions, see Oberguggenberger & Rosinger [chapter 7], or the global version of the classical Cauchy-Kovalevskaia theorem on analytic nonlinear PDEs, see Oberguggenberger & Rosinger and the literature cited there.

3) As seen from the proof of Proposition 2.2., the functions U_ϵ can in fact be chosen as piecewise polynomials in $x \in \mathbb{R}^n$.

Let us now note that there is an obvious ambiguity with the piecewise smooth functions in $\mathcal{C}^l_{nd}(\Omega)$ in (1.7). Indeed, given any such function $u \in \mathcal{C}^l_{nd}(\Omega)$, the corresponding closed, nowhere dense set Γ cannot be defined uniquely. Therefore, it is convenient to factor out this ambiguity, and this can be done easily as follows. Since $\mathcal{C}^0_{nd}(\Omega)$ is the largest of these spaces of functions, we shall do for this space the mentioned factoring out, by defining on it the *equivalence* relation $u \approx v$ for any two elements $u, v \in \mathcal{C}^0_{nd}(\Omega)$, as given by the condition

(2.17)
$$\exists\ \Gamma \subset \Omega\ \text{closed, nowhere dense}:$$
$$*)\ u,\ v \in \mathcal{C}^0(\Omega \setminus \Gamma)$$
$$**)\ u = v\ \text{on}\ \Omega \setminus \Gamma$$

It is easy to see that $\approx$ defined above is indeed an equivalence relation, since the union of a finite number of closed and nowhere dense subsets is again closed and nowhere dense. Now we can eliminate the mentioned ambiguity by going to the *quotient* space

(2.18) $\quad \mathcal{M}^0(\Omega) = \mathcal{C}^0_{nd}(\Omega)/\approx$

and in view of (2.8), (2.15), we define for any two elements $G, H \in \mathcal{M}^0(\Omega)$ the partial order $G \leq H$, by

(2.19)
$$\exists\ g \in G,\ h \in H,\ \Gamma \subset \Omega\ \text{closed, nowhere dense}:$$
$$*)\ g,\ h \in \mathcal{C}^0(\Omega \setminus \Gamma)$$
$$**)\ g \leq h\ \text{on}\ \Omega \setminus \Gamma$$

Let us now denote by

(2.20) $\quad (\mathcal{M}^0(\Omega)^\#, \leq)$

the Dedekind order completion due to MacNeille of the partially ordered space $(\mathcal{M}^0(\Omega), \leq)$ which was defined in (2.18), (2.19). Then this space $\mathcal{M}^0(\Omega)^\#$ in (2.20) is *order complete*, and we have the *order isomorphical embedding*, see Oberguggenberger & Rosinger [Appendix]

(2.21) $\quad \mathcal{M}^0(\Omega) \ni G \longmapsto\ < G\,] \in \mathcal{M}^0(\Omega)^\#$

which also *preserves the infima and suprema*. Moreover, in view of the MacNeille construction, we can further extend the embedding (2.21) as follows

(2.22)
$$\begin{array}{ccccc}
& \text{o. i. e.} & & \text{id} & \\
\mathcal{M}^0(\Omega) & \longrightarrow & \mathcal{M}^0(\Omega)^\# & \longrightarrow & \mathcal{P}(\mathcal{M}^0(\Omega)) \\
G & \longrightarrow & <G\,] & \longrightarrow & <G\,]
\end{array}$$

In order to obtain the full situation with respect to the range of the nonlinear partial differential operators $T(x, D)$, we note that $\mathcal{C}^0(\Omega) \subseteq \mathcal{C}^0_{nd}(\Omega)$, and we have the *order isomorphical embedding*

(2.23) $\quad \mathcal{C}^0(\Omega) \ni g \longmapsto\ G \in \mathcal{M}^0(\Omega)$

where G is the $\approx$ equivalence class of g. Furthermore, the partial order $\leq$ on $\mathcal{M}^0(\Omega)$ induces on $\mathcal{C}^0(\Omega)$ through this embedding the usual point-wise order relation of functions, namely, $g \leq h$, if and only if $g(x) \leq h(x)$, for $x \in \Omega$.

Let us now recall that our main interest is the construction of the commutative diagrams (1.9). In this regard, having constructed in (2.18) - (2.20), respectively, the spaces $\mathcal{M}^0(\Omega)$ and $\mathcal{M}^0(\Omega)^\#$ and their partial orders, the next step is to construct the partially ordered spaces $\mathcal{M}_T^m(\Omega)$ and $\mathcal{M}_T^m(\Omega)^\#$. For that purpose we start with (1.11), namely

(2.24) $\quad T(x, D) : \mathcal{C}_{nd}^m(\Omega) \longrightarrow \mathcal{C}_{nd}^0(\Omega)$

As mentioned, we shall solve the nonlinear PDEs in (1.1) by extending through order completion this mapping in (2.24), and we do so by constructing the commutative diagrams in (1.9). And at this stage we are now in the position to start doing so step by step. Let us note first that if $u \in \mathcal{C}^m(\Omega \setminus \Gamma)$, where $\Gamma \subset \Omega$ is any given closed, nowhere dense subset, then we also have

(2.25) $\quad T(x, D)u \in \mathcal{C}^0(\Omega \setminus \Gamma)$

This means that the singularity subsets Γ do *not* increase by the application of the nonlinear partial differential operators $T(x, D)$. However, as before with $\mathcal{C}_{nd}^0(\Omega)$, the ambiguity about associating such singularity sets to functions in $\mathcal{C}_{nd}^m(\Omega)$ remains. Therefore, in view of (2.25), we shall define the *equivalence* relation $u \approx_T v$, for any two functions $u, v \in \mathcal{C}_{nd}^m(\Omega)$, by the condition

(2.26) $\quad T(x, D)u \approx T(x, D)v$

which uses the equivalence relation $\approx$ given in (2.17), and in addition, it also depends on the nonlinear partial differential operator $T(x, D)$. In fact, this equivalence relation $\approx_T$ on $\mathcal{C}_{nd}^m(\Omega)$ is what is called the *pull-back* through the mapping $T(x, D)$ in (2.24) of the equivalence relation $\approx$ on $\mathcal{C}_{nd}^0(\Omega)$.

Let us now define the quotient space

(2.27) $\quad \mathcal{M}_T^m(\Omega) = \mathcal{C}_{nd}^m(\Omega) / \approx_T$

in which case the mapping (2.24) generates canonically the *injective* mapping

(2.28) $\quad T : \mathcal{M}_T^m(\Omega) \longrightarrow \mathcal{M}^0(\Omega)$

defined by $T(U) = G$, where G is the unique $\approx$ equivalence class in $\mathcal{M}^0(\Omega)$ of any of the $T(x, D)u$, where u belongs to the $\approx_T$ equivalence class U in $\mathcal{M}_T^m(\Omega)$.

At last, we can define the partial order $\leq_T$ on $\mathcal{M}_T^m(\Omega)$ as the *pull-back* through the mapping T in (2.28) of the partial order $\leq$ in (2.19) on $\mathcal{M}^0(\Omega)$, that is, for $U, V \in \mathcal{M}_T^m(\Omega)$, we have $U \leq_T V$, if and only if

(2.29) $\quad TU \leq TV$

In this way we obtain the partially ordered set $(\mathcal{M}_T^m(\Omega), \leq_T)$ giving the desired order structure on the domain $\mathcal{M}_T^m(\Omega)$ of T, which is the mapping in (2.28) that corresponds now to our nonlinear partial differential operator $T(x, D)$ in (1.2), (1.3), or more precisely, in (2.24). It is obvious in view of (2.29) that the injective mapping T in (2.28) is also an *order isomorphical embedding*.

So far, we have in this way obtained the top commutative rectangle in (1.9).

Applying now to $(\mathcal{M}_T^m(\Omega), \leq_T)$ the Dedekind order completion of MacNeille, we obtain

(2.30) $\quad (\mathcal{M}_T^m(\Omega)^\#, \leq_T)$

which is *order complete*, and in addition, similar with (2.21), we also have the *order isomorphical embedding*

(2.31) $\quad \mathcal{M}_T^m(\Omega) \ni U \longmapsto\ <U\,] \in \mathcal{M}_T^m(\Omega)^\#$

which *preserves the infima and suprema*. Also, similar with (2.22), we have

(2.32)
$$\begin{array}{ccccc} & \text{o. i. e.} & & \text{id} & \\ \mathcal{M}_T^m(\Omega) & \longrightarrow & \mathcal{M}_T^m(\Omega)^\# & \longrightarrow & \mathcal{P}(\mathcal{M}_T^m(\Omega)) \\ U & \longrightarrow & <U\,] & \longrightarrow & <U\,] \end{array}$$

And now all that remains is to define $T^\#$ in (1.9). In view of (2.32), however, this can be done in a standard manner following from the MacNeille order completion, see Oberguggenberger & Rosinger [Appendix]. Consequently, one obtains the *order isomorphical embedding*

(2.33) $\quad T^\# : \mathcal{M}_T^m(\Omega))^\# \longrightarrow \mathcal{M}^0(\Omega)^\#$

which also *preserves the infima and the suprema*. In more detail, we have the following commutative diagram

(2.34)
$$\begin{array}{ccc} \mathcal{M}_T^m(\Omega) \ni U & \xrightarrow{\;T\;} & T(U) \in \mathcal{M}^0(\Omega) \\ \downarrow & & \downarrow \\ \mathcal{M}_T^m(\Omega)^\# \ni\ <U\,] & \xrightarrow{\;T^\#\;} & \begin{array}{l} T^\#(<U\,]) = \\ =\ <T(U)\,] \in \mathcal{M}^0(\Omega) \end{array} \end{array}$$

In this way we have indeed obtained the whole of the commutative diagram (1.9), which we shall present now in the form seen next. Here "sur" and "inj" mean mappings which are surjective and injective, respectively, while as before, "o. i. e." means order isomorphic embedding, and "o. i." stands for order isomorphism. The dotted arrows "<- - - - - - -" mean the "pull-back" through which the respective structures were defined

(2.35)
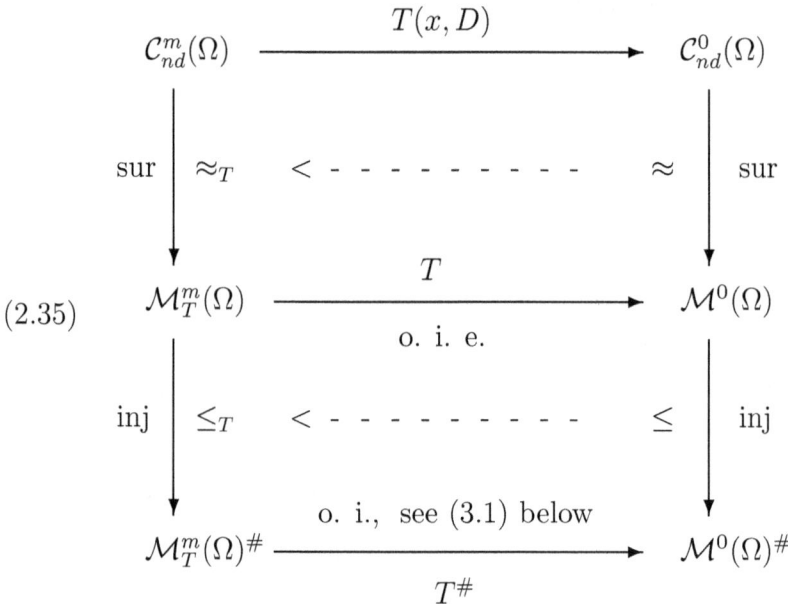

3. General existence result

One of the typical *main existence results* concerning the solutions of the nonlinear PDEs in (1.1) is presented in the following theorem, see Oberguggenberger & Rosinger [38-64] for a proof

Theorem 3.1.

(3.1) $T^\# (\mathcal{M}_T^m(\Omega)^\#) = \mathcal{M}^0(\Omega)^\#$

This means that, given the nonlinear PDEs in (1.1), for every right hand term $f \in \mathcal{M}^0(\Omega)^\#$, there exists a generalized solution $U \in \mathcal{M}_T^m(\Omega)^\#$, satisfying the relation $T^\# U = f$, according to the extension in (1.9).

As seen in Oberguggenberger & Rosinger [74-93], the space $\mathcal{M}^0(\Omega)^\#$ in which the right hand terms f of the nonlinear PDEs in (1.1) can range - and which now are solved by Theorem 3.1. - contains a large amount of *discontinuous* function on Ω. Certainly, in view of (2.18), $\mathcal{M}^0(\Omega)^\#$ contains all the *piecewise discontinuous* functions in $\mathcal{C}^0_{nd}(\Omega)$. What is particularly important to note is that, in view of (3.1), a variety of linear and nonlinear PDEs can be solved, in spite of the fact that the respective PDEs are known *not* to have solutions in distributions. Among them is the celebrated 1957 Hans Lewy example, see Obrguggenberger & Rosinger [chap. 6, 8].

In this regard, it was in Oberguggenberger & Rosinger that this Hans Lewy example of a PDE not solvable in distribution was nevertheless solved for the first time through the method of order completion.

The *coherence* between the solutions obtained in (3.1) and the usual classical solutions, whenever the nonlinear PDEs in (1.1) may have the latter, follows easily from the commutative diagram (2.35). In other words, whenever the nonlinear PDEs in (1.1) happen to have classical solutions $U \in \mathcal{C}^m(\Omega)$, then they are also generalized solution in the sense of (3.1).

Finally, it is important to note that the above existence result in (3.1) can easily be extended to *systems* of nonlinear PDEs of the general form in (1.1), see Oberguggenberger & Rosinger [chap. 8-11].

4. Initial and/or boundary value problems and constitutive relations

One of the significant advantages of the order completion method in solving systems of nonlinear PDEs of the general form in (1.1) comes from the *ease* initial and/or boundary value problems associated with such equations can be solved. This is in strong contradistinction with the variety of functional analytic methods of solution where considerable difficulties arise related to the need to restrict distributions or generalized functions to lower dimensional manifolds. Indeed, such operations of restriction are typically ill-defined.

On the other hand, when using the method of order completion, the issue of satisfying the initial and/or boundary values can be *decoupled* from the issue of the existence of solutions. Indeed, satisfying the initial and/or boundary values can be dealt with *first* and *separately from* the issue of proving the existence of solutions.

Details in this regard can be found in Oberguggenberger & Rosinger [chap. 8, 11]. And the reason behind that rather surprising ease the order completion method exhibits when dealing with initial and/or boundary value problems comes from a fact seen next, in section 5.

In Fluid Dynamics, and in general, Continuum Mechanics, a critical role is played by *constitutive equations*, see Rajagopal & Wineman, Rajagopal, or Rajagopal & Srinivasa.
The usual functional analytic methods can - if at all - deal with such constitutive equations in no less a difficult manner than they can do with initial and boundary value conditions. A regrettable consequence of these considerable difficulties is the failure so far of functional analytic methods to approach in any significant, let alone, systematic manner, the issue of such critically important constitutive equations. Here again, the order completion method proves its advantage by being able to deal as well with constitutive equations. This is simply a consequence of the fact that, as mentioned next in section 5, the order completion method can solve equations which are far more general than the linear or nonlinear systems of PDEs, or the constitutive equations.

5. An abstract existence result

A better understanding of the power underlying the order completion method in solving very large classes of equations, classes *far beyond* any nonlinear systems of PDEs, can be obtained from the following rather abstract existence result, see Oberguggenberger & Rosinger [chap. 9].
Let X be any set, and let $(Y, \leq)$ be any partially ordered set which has no minimum or maximum. Further, let

(5.1) $\quad T : X \longrightarrow Y$

be any given mapping. The problem we consider is to find a solution $A \in X$ for the equation

(5.2) $\quad T(A) = F$

for any given $F \in Y$. The answer is obtained as follows. Similar with the construction of the commutative diagrams (1.9) and (2.35), one can construct commutative diagrams

(5.3)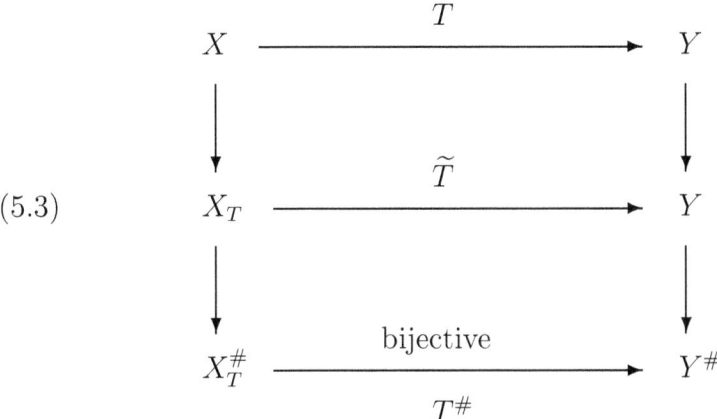

And then the following result on the existence of solutions holds

Theorem 5.1.

For any given $F \in Y^{\#}$, the equation

(5.4) $\quad T^{\#}(A) = F$

has a solution $A \in X_T^{\#}$, if and only if

(5.5) $\quad \sup\nolimits_{Y^{\#}} \{\, T^{\#}(U) \mid U \in X_T^{\#},\ T^{\#}(U) \subseteq F \,\} =$
$\qquad\quad = \inf\nolimits_{Y^{\#}} \{\, T^{\#}(V) \mid V \in X_T^{\#},\ F \subseteq T^{\#}(V) \,\}$

The significant generality of the above existence result allows, among others, the *separation* mentioned in section 4, between first satisfying the initial and/or boundary value conditions, and then, second, proving the existence of solutions in the case of general nonlinear systems of PDEs of the form in (1.1). Indeed, by first imposing the initial and/or boundary values, one is in fact defining the set X in (5.1). And as seen above, that can be done without any restrictions. Subsequently, condition (5.5) is both necessary and sufficient for the existence of a generalized solution $A \in X_T^{\#}$.

6. The Hausdorff continuity of solutions

The major *novelty* in this paper is about the fact that the solutions $U \in \mathcal{M}_T^m(\Omega)^{\#}$ of systems of nonlinear PDEs of type (1.1), obtained according to the procedure in Theorem 3.1., and of its generalizations can now be assimilated with *Hausdorff continuous functions* in $\mathbb{H}(\Omega)$. In fact, as seen in (A.12) in the Appendix, the mentioned solutions can be assimilated with *nearly finite Hausdorff continuous functions.*

7. Final comments

A further advantage of the order completion method is that one is *not* limited to consider in (1.9) and (2.35) only the *pull-back* partial order $\leq_T$ generated by the partial differential operators $T(x, D)$ on $\mathcal{M}_T^m(\Omega)$. Indeed, as seen in Oberguggenberger & Rosinger [chap. 13], a large variety of other partial orders on $\mathcal{M}_T^m(\Omega)$ can still secure existence theorems similar to Theorem 3.1.

As for the use of pull-back orders, it is important to note that there exists a certain analogy with functional analytic methods for solving PDEs. Indeed, in such methods, the topologies considered on the domains of the partial differential operators $T(x, D)$ are but pull-backs through these operators of suitable topologies on their ranges.

Details about such facts, and in general, about certain similarities between the order completion method and the usual functional analytic ones in solving PDEs can be found in Oberguggenberger & Rosinger [chap. 12].

The results in this paper invite a *comparison* with the *customary perception* regarding the solution of linear or nonlinear PDEs. Typical for that perception are the following to recent citations.

The 2004 edition of the Springer Universitext book "Lectures on PDEs" by V I Arnold, starts on page 1 with the statement :

> "In contrast to ordinary differential equations, there is *no unified theory* of partial differential equations. Some equations have their own theories, while others have no theory at all. The reason for this complexity is a more complicated geometry ..." (italics added)

Similarly, the 1998 edition of the book "Partial Differential Equations" by L C Evans, starts his Examples on page 3 with the statement :

> "There is no general theory known concerning the solvability of all partial differential equations. Such a theory is *extremely unlikely* to exist, given the rich variety of physical, geometric, and probabilistic phenomena which can be

modelled by PDE. Instead, research focuses on various particular partial differential equations ..." (italics added)

Appendix : Definition and Properties of Hausdorff-Continuous Functions

The Hausdorff continuous functions are not unlike the usual real valued continuous functions. For instance, they assume real values on a dense subset of their domain of definition and are completely determined by the values on this subset. However, these functions may also assume *interval* values on a certain subset of their domain of definition. Hence the concept of Hausdorff continuity is formulated within the realm of *interval valued functions*. We shall deal in this Appendix with functions whose values can be not only usual real numbers but also *extended* real numbers, that is, elements in $\overline{\mathbb{R}} = \mathbb{R} \cup \{-\infty, +\infty\}$. Moreover, as mentioned it proves to be convenient to allow the values of the functions to be not only numbers in $\overline{\mathbb{R}}$, but also *closed intervals* of such numbers, namely, $[a, b] \subseteq \overline{\mathbb{R}}$, with $a, b \in \overline{\mathbb{R}}$, $a \leq b$.

Towards the end of the 19-th century, Baire brought in the concepts of *lower* and *upper semi-continuous* functions, when dealing with non-smooth real valued functions. And in effect, he associated with each real valued function f, *two* other real, or extended real valued functions $I(f)$ and $S(f)$, with $I(f) \leq f \leq S(f)$, which proved to be particularly helpful, see (A.6), (A.7). However, following the prevailing mentality at the time, each of these three functions was considered separately and as being a single valued function.

As it turns out on the other hand, by considering *interval valued* functions, such as for instance $F(f) = [I(f), S(f)]$, one can significantly improve on the understanding and handling of non-continuous functions.

The study of interval valued functions can, among others, show that the particular case of functions which have values given by one single number is appropriate for continuous functions only. On the other hand, non-continuous functions are much better described by suitably associated interval valued functions.

Indeed, in the case of functions f which are *not* continuous, a much better description can be obtained by considering them given by a *pair* of usual point valued functions, namely $f = [\,\underline{f}, \overline{f}\,]$, thus leading to interval valued functions, according to $f(x) = [\underline{f}(x), \overline{f}(x)] \subseteq \overline{\mathbb{R}}$, for $x \in \Omega$. And then, a natural class which replaces, and also extends, the usual point valued continuous functions is that of *Hausdorff-continuous* interval valued functions, see below Definition A1. The distinctive and *essential* feature of these Hausdorff-continuous functions $f = [\underline{f}, \overline{f}]$ is a condition of *minimality* with respect to the *gap* between $\underline{f}$ and $\overline{f}$, with the further requirement that $\underline{f}$ be lower semi-continuous, and $\overline{f}$ be upper semi-continuous.

The interest in more recent times in interval valued functions comes from a number of branches of mathematics, such as approximation theory, Sendov, and numerical analysis, Kraemer.

Most of the results on interval valued functions presented in this Appendix have, however, been developed by R Anguelov. This was done in view of the usefulness of such functions in several branches of mathematics, see Anguelov [1,2], Anguelov & Markov, Anguelov et.al [1,2], Anguelov & Minani, Anguelov & Rosinger [1-3]. Here, owing to restriction of space, we shall only present a minimal amount of them, needed in order to support section 6 above. For the full details in this regard, including proofs, see Appendix 2 in Anguelov & Rosinger [3].

As mentioned, the class of interval valued functions of special interest here is that of *Hausdorff-continuous*, or in short, *H-continuous* functions. As it turns out they enjoy the *minimality* property (A.8) with respect to their *graph completion*, see (A.7), and that allows an effective interplay between Analysis and Topology, with the latter involving both the domain and the range of the functions dealt with. Let

(A.1) $\overline{\mathbb{IR}} = \{[\underline{a}, \overline{a}] \mid \underline{a}, \overline{a} \in \overline{\mathbb{R}} = \mathbb{R} \cup \{-\infty, +\infty\}, \underline{a} \leq \overline{a}\}$

be the set of all finite or infinite closed intervals. The functions which we consider can be defined on arbitrary topological spaces Ω. For the purposes of the nonlinear PDEs studied in this paper, however, it will be sufficient to assume that $\Omega \subseteq \mathbb{R}^n$ are arbitrary open subsets. Let us now consider the set of interval valued functions

(A.2) $\mathbb{A}(\Omega) = \{f : \Omega \longrightarrow \overline{\mathbb{IR}}\}$

By identifying the point $a \in \overline{\mathbb{R}}$ with the degenerate interval $[a, a] \in \overline{\mathbb{IR}}$, we consider $\overline{\mathbb{R}}$ as a subset of $\overline{\mathbb{IR}}$. In this way $\mathbb{A}(\Omega)$ will contain the set of functions with extended real values, namely

(A.3) $\mathcal{A}(\Omega) = \{f : \Omega \longrightarrow \overline{\mathbb{R}}\} \subseteq \mathbb{A}(\Omega)$

We define a partial order $\leq$ on $\overline{\mathbb{IR}}$ by

(A.4) $[\underline{a}, \overline{a}] \leq [\underline{b}, \overline{b}] \iff \underline{a} \leq \underline{b},\ \overline{a} \leq \overline{b}$

Now on $\mathbb{A}(\Omega)$ we define the partial order induced by (A.4) in the usual point-wise way, namely, for $f,\ g \in \mathbb{A}(\Omega)$, we have

(A.5) $f \leq g \iff f(x) \leq g(x),\ x \in \Omega$

Clearly, when restricted to $\mathcal{A}(\Omega)$, the above partial order on $\mathbb{A}(\Omega)$ reduces to the usual one among point valued functions.

Let $f \in \mathbb{A}(\Omega)$. For every $x \in \Omega$, the value of f is an interval, namely, $f(x) = [\,\underline{f}(x),\ \overline{f}(x)\,]$, with $\underline{f}(x),\ \overline{f}(x) \in \overline{\mathbb{R}}$, $\underline{f}(x) \leq \overline{f}(x)$. Hence, every function $f \in \mathbb{A}(\Omega)$ can be written in the form $f = [\,\underline{f},\ \overline{f}\,]$, with $\underline{f},\ \overline{f} \in \mathcal{A}(\Omega)$, $\underline{f} \leq f \leq \overline{f}$, and $f \in \mathcal{A}(\Omega) \iff \underline{f} = f = \overline{f}$.

In the particular case of functions in $\mathcal{A}(\Omega)$, that is, with extended real, but point, and not nondegenerate interval values, a number of basic results were obtained already in Baire, see also Nicolescu for a more recent detailed presentation. The rest of the more general results concerning functions in $\mathbb{A}(\Omega)$, that is, with values finite or infinite closed intervals, were developed for the first time in the above cited works of Anguelov . The few such earlier results were obtained in Sendov, where the particular instance of $\Omega \subseteq \mathbb{R}$ was dealt with.

For $x \in \Omega$, we denote by $B_\delta(x)$ the open ball of radius δ centered at x. Let us consider any *dense* subset $D \subseteq \Omega$, and associate with it the pair of mappings $I(D,\Omega,.),\ S(D,\Omega,.) : \mathbb{A}(\Omega) \to \mathcal{A}(\Omega)$, called *lower* and *upper Baire operators*, respectively, where for every function $f \in \mathbb{A}(\Omega)$ and $x \in \Omega$, we define

(A.6) $\begin{aligned} I(D,\Omega,f)(x) &= \sup_{\delta > 0}\ \inf\ \{\,z \in f(y)\ |\ y \in B_\delta(x) \cap D\,\} \\ S(D,\Omega,f)(x) &= \inf_{\delta > 0}\ \sup\ \{\,z \in f(y)\ |\ y \in B_\delta(x) \cap D\,\} \end{aligned}$

In Baire, these two operators were considered and studied in the particular case of functions $f \in \mathcal{A}(\Omega)$ and when $D = \Omega$, see also Nicolescu. In view of the main interest here in *interval valued* functions $f \in \mathbb{A}(\Omega)$, it is useful to consider as well the following third mapping, namely, $F(D,\Omega,f) : \mathbb{A}(\Omega) \to \mathbb{A}(\Omega)$, defined for $f \in \mathbb{A}(\Omega)$ by

(A.7) $F(D,\Omega,f)(x) = [\,I(D,\Omega,f)(x),\ S(D,\Omega,f)(x)\,],\quad x \in \Omega,$

and called the *graph completion operator*. In case $D = \Omega$, we use the simpler notations $I(\Omega,\Omega,f) = I(f)$, $S(\Omega,\Omega,f) = S(f)$ and $F(\Omega,\Omega,f) = F(f)$.

The next definition was given in Sendov in the case of $\Omega \subseteq \mathbb{R}$, however, it can obviously be extended to any topological space Ω.

Definition A.1

A function $f \in \mathbb{A}(\Omega)$ is called *Hausdorff-continuous*, or in short, *H-continuous*, if and only if for every function $g \in \mathbb{A}(\Omega)$, we have satisfied the following *minimality* condition on f

(A.8) $\quad g(x) \subseteq f(x), \quad x \in \Omega \quad \Longrightarrow \quad F(g) = f$

We shall denote by $\mathbb{H}(\Omega)$ the set of all Hausdorff-continuous interval valued functions on Ω.

Definition A2

A function $f \in \mathbb{H}(\Omega)$ is called *nearly finite*, if and only if there exists an open and dense subset $D \subseteq \Omega$, such that

(A.9) $\quad f(x) \in \overline{\mathbb{IR}} \quad$ is a finite interval for $\quad x \in D$

We denote by $\mathbb{H}_{nf}(\Omega)$ the set of nearly finite H-continuous functions $f \in \mathbb{A}(\Omega)$.

Regarding the *regularity* properties of solutions of general nonlinear systems of PDEs of the form in (1.1), a crucial role is played by the following mapping

(A.10) $\quad F_0 : \mathcal{C}^0_{nd}(\Omega) \ni u \longmapsto F(\Omega \setminus \Gamma, \Omega, u) \in \mathbb{H}_{nf}(\Omega)$

where we recall that, according to (1.7), for every $u \in \mathcal{C}^0_{nd}(\Omega)$, there exists a closed, nowhere dense subset $\Gamma \subset \Omega$, such that $u \in \mathcal{C}^0(\Omega \setminus \Gamma)$, hence in view of (A.7), $F(\Omega \setminus \Gamma, \Omega, u)$ is well defined. The fact that such a Γ need not be unique, does not affect the above definition, see Anguelov & Rosinger [3].

The following theorem shows that the images of two functions in $\mathcal{C}_{nd}(\Omega)$ under the mapping F_0 in (A.10) are the same, if and only if these functions are equivalent with respect to the equivalence relation (2.17).

Theorem A.1

Let $u, v \in \mathcal{C}_{nd}(\Omega)$. Then $F_0(u) = F_0(v) \iff u \approx v$

In view of (2.18), (A.10) and the above theorem now we can define a mapping

(A.11) $\quad \widetilde{F}_0 : \mathcal{M}^0(\Omega) \longrightarrow \mathbb{H}_{nf}(\Omega)$

in the following way. Let $u \in U \in \mathcal{M}^0(\Omega)$, then $\widetilde{F}_0(U) = F(u)$. It is easy to see that the definition of $\widetilde{F}_0(U)$ does not depend on the particular representative $u \in U$ of the equivalence class U.

Theorem A.2

The mapping $\widetilde{F}_0 : \mathcal{M}^0(\Omega) \longrightarrow \mathbb{H}_{nf}(\Omega)$ defined in (A.11) is an *order isomorphic embedding* with respect to the order relation (2.19) on $\mathcal{M}^0(\Omega)$ and the order relation induced by (A.5) on $\mathbb{H}_{nf}(\Omega)$. Namely, for any $U, V \in \mathcal{M}^0(\Omega)$, we have

$$U \leq V \iff \widetilde{F}_0(U) \leq \widetilde{F}_0(V)$$

Finally, we also have

Theorem A.3

The set $\mathbb{H}_{nf}(\Omega)$ is Dedekind order complete with respect to the partial order induced on it by (A.5).

Let $g \in \mathbb{H}_{nf}(\Omega)$. Then there exists a subset $\mathcal{G} \subseteq \mathcal{M}^0(\Omega)$ such that

$$g = \sup \widetilde{F}_0(\mathcal{G}) = \sup\{\widetilde{F}_0(G) \mid G \in \mathcal{G}\}$$

This theorem shows that $\mathbb{H}_{nf}(\Omega)$ is the *smallest* Dedekind order complete subset of $\mathbb{H}(\Omega)$ which contains the image of $\mathcal{M}^0(\Omega)$ under the order isomorphical embedding $\widetilde{F}_0$. Hence it is *order isomorphic* to the Dedekind order completion $\mathcal{M}^0(\Omega)^{\#}$ of $\mathcal{M}^0(\Omega)$. In this way we obtain the commutative diagram, where $\widetilde{F}_0^{\#}$ denotes the order isomorphism from $\mathcal{M}^0(\Omega)^{\#}$ to $\mathbb{H}_{nf}(\Omega)$, namely

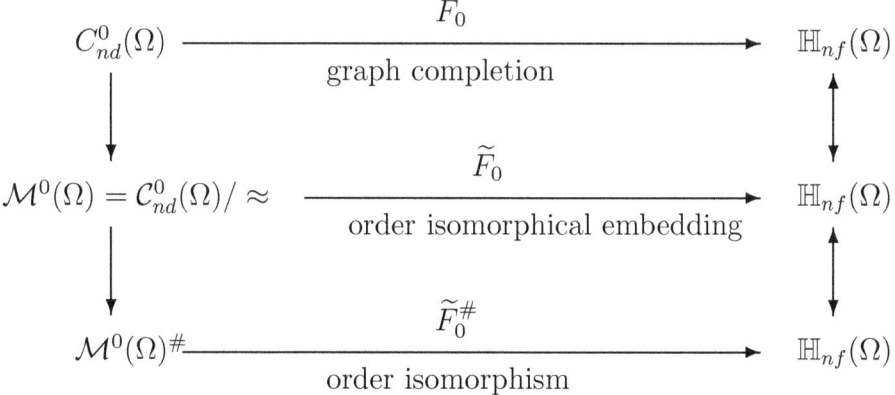

Now we can bring together the above diagram with the one in (2.35) and obtain the two successive *order isomorphisms*

$$(A.12) \quad \mathcal{M}_T^m(\Omega))_T^{\#} \xrightarrow{T^{\#}} \mathcal{M}^0(\Omega)^{\#} \xrightarrow{\widetilde{F}_0^{\#}} \mathbb{H}_{nf}(\Omega)$$

It follows therefore that the set $\mathcal{M}_T^m(\Omega))_T^{\#}$ in which the *solutions* of the general nonlinear systems of PDEs of the form in (1.1) are found is mapped by the *bijection* $\widetilde{F}_0^\# \circ T^\#$ onto the set $\mathbb{H}_{nf}(\Omega)$ of all *nearly finite Hausdorff continuous functions*. Since both these mappings are order isomorphisms, the set $\mathcal{M}_T^m(\Omega))_T^\#$ is order isomorphic with the set $\mathbb{H}_{nf}(\Omega)$.

Hence, the solutions of the general nonlinear systems of PDEs of the form in (1.1), which are obtained through the order completion method, can always be assimilated with nearly finite Hausdorff continuous functions.

This is the argument supporting section 6 above.

For the sake of further clarification, let us turn now in the remaining part of this Appendix to some of the issues concerning the *discontinuities* of Hausdorff-continuous functions. Arbitrary interval valued functions $f = [\,\underline{f},\,\overline{f}\,] \in \mathbb{A}(\Omega)$ can exhibit a variety of types of discontinuities, and certainly not less so, than usual point valued functions $f \in \mathcal{A}(\Omega)$ do.

Hausdorff-continuous functions, although generalize usual point valued continuous functions, enjoy nevertheless a number of nontrivial *continuity* related properties. On the other hand, Hausdorff-continuous functions can have quite large sets of *discontinuities*. This shows that they do indeed form a larger class than the usual continuous functions, even if they still have important similar properties.

The fact that large enough sets of discontinuities can be present with Hausdorff-continuous functions allows for their use - as seen in this paper - in obtaining the *existence of non-classical solutions* for large classes of systems of nonlinear PDEs.

Consequently, and as mentioned, Hausdorff-continuous functions - precisely since they are *not* generalized functions - can be seen as *setting aside* to a certain extent the variety of distributional and other traditional generalized solutions of linear and nonlinear PDEs which have been obtained by functional analytic methods, or by the methods of the nonlinear algebraic theory listed by the AMS Subject Classification 2000, under 46F30.

Indeed, solving large classes of nonlinear PDEs through Hausdorff-continuous functions offers, among others, the following *double advantage* :

- one can bring in a significant *simplification* by avoiding the variety of usual functional analytic methods with their spaces of distributions or generalized functions, and

- one can obtain *universal regularity* results for solutions of large classes of systems of nonlinear PDEs.

Theorems A.4 and A.5 below show important properties of *Hausdorff-continuous* functions related to their sets of *discontinuities*. For every interval valued function $f = [\,\underline{f},\,\overline{f}\,] \in \mathbb{A}(\Omega)$, we denote by $\Gamma(f) = \{\,x \in \Omega \mid \underline{f}(x) < \overline{f}(x)\,\}$, which is the set of points $x \in \Omega$ where f assumes values $f(x) = [\,\underline{f}(x),\,\overline{f}(x)\,]$ that are *non-degenerate intervals*, and not merely points.

It follows that at points $x \in \Gamma(f)$, the interval valued function $f = [\,\underline{f},\,\overline{f}\,]$ *cannot* be continuous in the usual sense, since it is not a usual point valued function.

In the particular case of *Hausdorff-continuous* functions, this fact can further be clarified. Namely, given any point $x \in \Omega$, then

(A.13) $\quad \begin{array}{l} x \in \Gamma(f) \iff \underline{f} \text{ and } \overline{f} \text{ not continuous at } x \iff \\ \iff \underline{f} \text{ or } \overline{f} \text{ not continuous at } x \iff \underline{f}(x) < \overline{f}(x) \end{array}$

And now the basic result on the discontinuities of Hausdorff-continuous functions

Theorem A.4

Given any H-continuous function $f \in \mathbb{A}(\Omega)$. Then $\Gamma(f)$ is of *first Baire category* in Ω.

Let us further specify the structure of the discontinuity set $\Gamma(f)$. For $\epsilon > 0$, let us denote $\Gamma_\epsilon(f) = \{\,x \in \Omega \mid \overline{f}(x) - \underline{f}(x) \geq \epsilon\,\}$. Then clearly $\Gamma(f) = \bigcup_{\epsilon > 0} \Gamma_\epsilon(f) = \bigcup_{n \geq 1} \Gamma_{1/n}(f)$.

The next theorem gives a further insight into the structure of the discontinuity set $\Gamma(f)$ of Hausdorff-continuous functions $f \in \mathbb{A}(\Omega)$.

Theorem A.5

If the function $f \in \mathbb{A}(\Omega)$ is H-continuous, then for every $\epsilon > 0$, the set $\Gamma_\epsilon(f)$ is *closed* and *nowhere dense* in Ω.

An important *similarity* between usual continuous, and on the other hand, Hausdorff-continuous functions is that both of them are determined *uniquely* if they are known on a *dense* subset of their domains of definition. This property comes in spite of the fact that, as seen in the previous section, Hausdorff-continuous functions can have discontinuities on sets of first Baire category, and such sets can have arbitrary large positive Lebesgue measure, see Oxtoby. Indeed, we have

Theorem A.6

Let $f = [\,\underline{f},\, \overline{f}\,]$, $g = [\,\underline{g},\, \overline{g}\,] \in \mathbb{A}(\Omega)$ be two H-continuous functions, and suppose given any dense subset $D \subseteq \Omega$. Then

$$f(x) = g(x),\ x \in D \implies f = g \text{ on } \Omega$$

The real line $\mathbb{R}$ is Dedekind order complete, but not order complete, while the extended real line $\overline{\mathbb{R}}$ is both Dedekind order complete and order complete.

Let us recall that a partially ordered set which is order complete will also be Dedekind order complete, but as seen above, not necessarily the other way round as well.

Typically, various spaces of real valued functions encountered in Analysis are neither Dedekind order complete, nor order complete, when considered with the natural point-wise partial order relation.

However, as we can see next, this situation changes when we deal with the set $\mathbb{H}(\Omega)$ of Hausdorff-continuous functions.

Theorem A.7 (Anguelov [2])

The set $\mathbb{H}(\Omega)$ of Hausdorff-continuous functions is order complete when considered with the partial order in (A.5).

The Dedekind order completeness of the space $\mathbb{H}(\Omega)$ of Hausdorff-continuous functions is a *nontrivial* property, in view of the various connections between the usual continuous, and on the other hand Hausdorff-continuous functions. Indeed, the space $\mathcal{C}(\Omega)$ of usual real valued continuous functions, which we have seen is strictly contained in $\mathbb{H}(\Omega)$, is well known *not* to be Dedekind order complete. On the other hand, once $\mathbb{H}(\Omega)$ proves to be Dedekind order complete, its order completeness follows easily from the fact that $\overline{\mathbb{R}}$ is order complete. Indeed, the smallest and largest elements in $\mathbb{H}(\Omega)$ are respectively the functions $\Omega \ni x \longmapsto -\infty$ and $\Omega \ni x \longmapsto +\infty$.

As is well known and shown by simple examples the spaces of real valued continuous functions $\mathcal{C}(\Omega)$ are *not* Dedekind order complete, thus, are not order complete either.

Since these spaces are partially ordered in a natural way, one can apply to them the MacNeille version of Dedekind order completion method, see Oberguggenberger & Rosinger [Appendix], Luxemburg & Zaanen, or Zaanen.

This however being a general construction based on Dedekind type cuts, it leaves open the question of the *nature* of the elements which are added to these spaces of continuous functions by the respective Dedekind order completion process.

A classical, 1950 result in this regard was obtained by Dilworth in the case of *bounded* and real valued continuous functions on Ω for arbitrary completely regular topological spaces Ω. Namely, the respective Dedekind order completion is given by all the normal upper semi-continuous functions on Ω. Regarding the Dedekind order completion of spaces $\mathcal{C}(\Omega)$ certain results were obtained in Mack & Johnson.

As seen in Anguelov [1], the Dedekind order completion of spaces $\mathcal{C}(\Omega)$ of real valued continuous functions was for the first time effectively constructed for a large class of topological spaces Ω. In this construction Hausdorff-continuous functions and some of their subspaces play a crucial role. Consequently, both the problem of the *completion* of the spaces $\mathcal{C}(\Omega)$, as well as that of the *structure* of the elements which are added to these spaces $\mathcal{C}(\Omega)$ through the respective completion find a convenient solution through the use of *interval valued* functions.

References

[1] Anguelov R [1] : Dedekind order completion of $C(X)$ by Hausdorff continuous functions. Quaestiones Mathematicae, Vol. 27, 2004, 153-170

[2] Anguelov R [2] : An introduction to some spaces of interval functions. arXiv:math.GM/0408013

[3] Anguelov R, Markov S : Extended segment analysis. Freiburger Intervall - Berichte 10, 1981, 1 - 63.

[4] Anguelov R, Markov S, Sendov B [1] : On the Normed Linear Space of Hausdorff Continuous Functions. Poceedings of the Fifth International Conference on "Large Scale Scientific Computations", June 6-10, 2005, Sozopol, Lecture Notes in Computer Science, Springer (to appear)

[5] Anguelov R, Markov S, Sendov B [2] : The Linear Space of Hausdorff Continuous Functions, Technical Report UPWT2004/4

[6] Anguelov R, Minani F : Interval Viscosity Solutions of Hamilton-Jacobi Equations. Technical Report UPWT 2005/3, University of Pretoria

[7] Anguelov R, Rosinger E E [1] : Dedekind order completion of $\mathcal{M}(\Omega)$ by Hausdorff continuous functions. (to appear)

[8] Anguelov R, Rosinger E E [2] : Hausdorff Continuous Solutions of Nonlinear PDEs through the Order Completion Method. Quaestiones Mathematicae (to appear), see arXiv:math.AP/0406517

[9] Anguelov R, Rosinger E E [3] : Solution of Nonlinear PDEs by Hausdorff Continuous Functions (to appear).

[10] Arnold V I : Lectures on PDEs. Springer Universitext, 2004

[11] Baire R : Lecons sur les Fonctions Discontinues. Collection Borel, Paris, 1905.

[12] Bardi M, Capuzzo-Dolcetta I : Optimal control and viscosity solutions of Hamilton-Jacobi-Bellman equations. Birkhäuser, Boston, Basel, Berlin, 1997.

[13] Birkhoff G : The Role of Order in Computing. In Moore R (Ed.) : Reliability in Computing. Academic Press, 1988, 357–378.

[14] Dilworth R P : The normal completion of the lattice of continuous functions. Trans. Amer. Math. Soc., 68, 1950, 427–438.

[15] Evans L C : Partial Differential Equations. AMS Graduate Studies in Mathematics, Vol. 19, 1998

[16] Forster O : Analysis 3, Integralrechnung in $\mathbb{R}^n$ mit Anwendungen. Friedr. Vieweg, Braunschweig, Wiesbaden, 1981

[17] Kraemer W, von Gudenberg J W (Eds) : Scientific Computing, Validated Numerics, Interval Methods. Kluwer, Dordrecht, 2001

[18] Lewy, H : An example of smooth linear partial differential equation without solutions. Ann. Math., vol. 66, no. 2, 1957, 155-158

[19] Luxemburg W A J, Zaanen A C : Riesz Spaces I. North Holland, Amsterdam, 1971.

[20] Mack J E, Johnson D G : The Dedekind completion of C(X). Pacif. J. Math., 20, 2, 1967, 231-243

[21] MacNeille H M : Partially ordered sets. Trans. AMS, vol. 42, 1937, 416-460

[22] Markov S : A nonstandard subtraction of intervals. Serdica. Vol. 3, 1977, 359-370

[23] Markov S : Calculus for interval functions of a real variable. Computing, Vol. 22, 1979, 325-337

[24] Markov S : Extended interval arithmetic involving infinite intervals. Mathematica Balkanica, Vol.6, 1992, 269-304.

[25] Nicolescu M : Analiză Matematică II. Editura Technică, București, 1958

[26] Oberguggenberger M B, Rosinger E E : Solution of Continuous Nonlinear PDEs through Order Completion. North-Holland Mathematics Studies, Vol. 181. North-Holland, Amsterdam, 1994

see also review MR 95k:35002

[27] Oxtoby J C : Measure and Category. Springer, New York, 1971

[28] Rajagopal K R, Wineman A S : On constitutive equations for branching of response with selectivity. International Journal of Nonlinear Mechanics, Vol. 15, 1980, 83-91

[29] Rajagopal K R : On implicit constitutive theories. Application of Mathematics, Vol. 28, No. 4, 2003, 279-319

[30] Rajagopal K R, Srinivasa A R : On thermo-mechanical restrictions of continua. Proc. R. Soc. London A Math., Vol. 460, 2004, 631-651

[31] Rosinger E E : Hausdorff continuous solutions of arbitrary continuous nonlinear PDEs through the order completion method. arXiv:math.AP/0405546

[32] Rosinger E E : Can there be a general nonlinear PDE theory for the existence of solutions ? arXiv:math.AP/0407026

[33] Rosinger E E, Rudolph M : Group invariance of global generalised solutions of nonlinear PDEs : A Dedekind order completion method. Lie Groups and their Applications, Vol. 1, No. 1, July-August 1994, 203-215

[34] Sendov B : Hausdorff Approximations. Kluwer, Dordrecht, 1990

[35] Tutschke W : Initial Value Problems in Classes of Generalized Analytic Functions. Springer, New York, 1989

[36] Zaanen A C : The universal completion of an Archimedean Riesz space. Indag. Math., 45, 4, 1983, 435-441

[37] Zaharov V : Functional characterization of absolute and Dedekind completion. Bull. Acad. Polon. Sci., 29, 5-6, 1981, 293-297

4. A Few General Results on Partial Orders

Extending Mappings between Posets, arXiv:mathh/0609234

Elemér E Rosinger

Department of Mathematics
and Applied Mathematics
University of Pretoria
Pretoria
0002 South Africa
eerosinger@hotmail.com

Abstract

A variety of possible extensions of mappings between posets to their Dedekind order completion is presented. One of such extensions has recently been used for solving large classes of nonlinear systems of partial differential equations with possibly associated initial and/or boundary value problems.

1. The General Setup

Let $(X, \leq)$ and $(Y, \leq)$ be two arbitrary posets and

(1.1) $\quad \varphi : X \longrightarrow Y$

any mapping between them. We shall be interested to set up *commutative diagrams*

(1.2)

where $X^\#$ and $Y^\#$ are the Dedekind order completions, [3,2,4], of X and Y, respectively, while the mappings

(1.3) $\quad \varphi^\diamond : X^\# \longrightarrow Y^\#$

are *extensions* of the given mapping in (1.1), in view of the commutativity of (1.2).

As we shall see, there are many natural ways to obtain extensions (1.3). One such way, see (A.26) - (A.28) and Proposition A.1 in the Appendix, has recently been used successfully in order to solve large classes of nonlinear systems of PDEs with possibly associated initial and/or boundary value problems, [4,1,5-7].

Several other earlier obtained results relating to posets and their Dedekind order completions, result needed in the sequel, are summarized in the Appendix.

In view of the main interest pursued being the solution of large classes of nonlinear systems of PDEs with possibly associated initial and/or boundary value problems, the sets X and Y are supposed to be infinite, since in the particular case when solving PDEs, they correspond to spaces of functions on Euclidean domains on which the respective PDEs are defined.

Furthermore, for the convenience of the Dedekind order completion method, [3], and without loss of generality, [3,2,4], we shall assume that the posets $(X, \leq)$ and $(Y, \leq)$ do not have minimum or maximum. Otherwise, these two posets can be arbitrary.

2. Constructing Extensions

It is quite natural to define the extension (1.3) as follows, see (A.7), (A.8)

$$(2.1) \quad \mathcal{P}(X) \ni A \longmapsto \varphi^\diamond(A) = (\varphi(A))^{ul} \in Y^\#$$

which enjoys the following two advantages :

- it has a larger domain of definition that required in (1.3), and furthermore

- it does not make use of the partial order on X.

This however, is precisely the definition of the mapping $\varphi^\#$ in (A.26) - (A.28) which, a mentioned, was given earlier in [4], and used in solving large classes of nonlinear systems of PDEs with possibly associated initial and/or boundary value problems, [4,1,5-7].

Consequently, we shall look for other possible extensions (1.3) which may similarly be natural.

Let us start by noting that the desired extended mapping $\varphi^\diamond$ in (1.2), (1.3) must be such that, given $A \subseteq X$, in order to obtain the corresponding $\varphi^\diamond(A) \subseteq Y$, one should not use more information than it is in the subset $\varphi(A) \subseteq Y$. This is precisely the reason $\varphi^\#$ was defined in the respective manner in (A.26), (A.27), see also (2.1) above.

And then, the way left for alternative definitions of $\varphi^\diamond$ is to try to use in the definition of $\varphi^\diamond(A) \subseteq Y$, with $A \subseteq X$, an amount of information which may possibly be *less* than that contained in $\varphi(A) \subseteq Y$.

A simplest way to do that is to define

(2.2) $\quad \tilde{\varphi} : \mathcal{P}(X) \longrightarrow Y^\#$

by

(2.3) $\quad \tilde{\varphi}(A) = \bigcap_{a \in A} (\varphi([a > \cap A]))^{ul}, \quad A \subseteq X$

This definition can obviously be generalized in the following manner. A mapping

(2.4) $\quad L : \mathcal{P}(X) \longrightarrow \mathcal{P}(X)$

is called *cofinal*, if and only if

$\forall \ A \subseteq X \ :$

(2.5) $\quad$ *) $\ L(A) \subseteq A$

$\quad\quad\quad$ * *) $\ L(A)$ is cofinal in A

Here we recall that a subset $B \subseteq A$ is *cofinal* in A, if and only if

(2.6) $\quad \forall \ a \in A \ : \ \exists \ b \in B \ : \ a \leq b$

And then we can define

(2.7) $\quad \varphi^L : \mathcal{P}(X) \longrightarrow \mathcal{P}(X)$

by

(2.8) $\quad \varphi^L(A) = \bigcap_{a \in L(A)} (\varphi([a > \cap A))^{ul}, \quad A \subseteq X$

This further suggests the following alternative possibility. Given $A \subseteq X$, instead of the subsets $[a > \cap A \subseteq A$, with $a \in A$, or $L(A) \subseteq A$, we can consider arbitrary subsets $B \subseteq A$.

However, in defining $\varphi^\diamond(A)$, one should not lose too much from the information in $\varphi(A)$. Thus there should be some *restriction* on what kind of subsets $B \subseteq A$ one is considering.

In this regard, and as above, a natural candidate is given by subsets $B \subseteq A$ which are *cofinal* in A. And then, we arrive at defining

(2.9) $\quad \overline{\varphi} : \mathcal{P}(X) \longrightarrow Y^\#$

by

(2.10) $\quad \overline{\varphi}(A) = \bigcap_{B \text{ cofinal in } A} (\varphi(B))^{ul}$

We note that, unlike $\widetilde{\varphi}$ and $\overline{\varphi}$ which are two possible definitions for $\varphi^\diamond(A)$, there can in general be *infinitely* many mappings φ^L, for any given pair of posets $(X, \leq)$ and $(Y, \leq)$.

We also note that in view of (A.11), one obtains

(2.11) $\quad \overline{\varphi}(A) \cup \widetilde{\varphi}(A) \cup \varphi^L(A) \subseteq \varphi^\#(A), \quad A \subseteq X$

3. Relations Among the Extended Mappings $\widetilde{\varphi}$, φ^L, $\overline{\varphi}$ and $\varphi^\#$

Proposition 3.1.

(3.1) $\quad \widetilde{\varphi} = \varphi^L$

for every *cofinal* mapping L in (2.4).

Proof

In view of (2.5), we have $L(A) \subseteq A$, thus (2.3), (2.8) yield the inclusion '$\subseteq$' in (3.1).

For the converse inclusion '$\supseteq$' in (3.1), we recall that $L(A)$ is cofinal in A, see (2.5). Hence for every $a \in A$, there exists $a' \in L(A)$, such that $a \leq a'$. Consequently, we have

$$[a' > \cap A \subseteq [a > \cap A$$

thus

$$\varphi([a' > \cap A) \subseteq \varphi([a > \cap A)$$

and then (A.11) implies

$$(\varphi([a' > \cap A))^{ul} \subseteq (\varphi([a > \cap A))^{ul}$$

and the proof of (3.1) is completed.

Proposition 3.2.

Let $A \subseteq X$ be *directed*, then

(3.2) $\quad \overline{\varphi}(A) \subseteq \widetilde{\varphi}(A)$

Here we recall that $A \subseteq X$ is *directed*, if and only if

(3.3) $\quad \forall\ a,\ a' \in A : \exists\ a'' \in A : a \leq a'',\ a' \leq a''$

Proof

In view of (2.3), let $a \in A$, then $B = [a > \cap A$ is cofinal in A, since A is directed. Hence (2.10) gives the inclusion in (3.2).

Proposition 3.3.

If the mapping φ in (1.1) is *increasing*, then

(3.4) $\quad \overline{\varphi} = \varphi^{\#}$

Proof

We shall show that

(3.5) $\quad (\varphi(A))^{ul} = (\varphi(B))^{ul}$

for every $B \subseteq A$, with B cofinal in A. Indeed, for every $a \in A$, there exists $b \in B$, such that $a \leq b$. But φ is increasing, hence $\varphi(a) \leq \varphi(b)$, which means

$$[\varphi(b) > \subseteq [\varphi(a) >$$

thus in view of (A.2), we obtain

$$(\varphi(B))^u \subseteq (\varphi(A))^u$$

But $B \subseteq A$ and (A.11) always imply

$$(\varphi(A))^u \subseteq (\varphi(B))^u$$

Hence in our case (3.5) does indeed hold. And then (3.4) follows from (2.10) and (A.27).

Corollary 3.1

If the mapping φ in (1.1) is *increasing*, then

(3.6) $\quad \overline{\varphi}(A) = \widetilde{\varphi}(A) = \varphi^L(A) = \varphi^{\#}(A)$

for every *directed* $A \subseteq X$.

4. Extension Diagrams

Let us return now to the initial main problem, namely, to construct extensions (1.2) for arbitrary mappings (1.1) by using Dedekind order completions.

Theorem 4.1

Let φ in (1.1) be an arbitrary mapping, then the following two diagrams are commutative

(4.1)
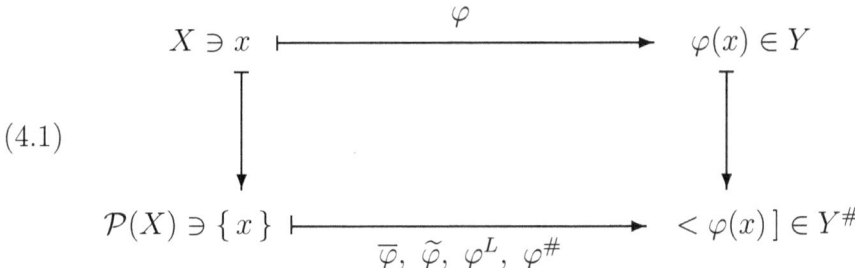

and

(4.2)
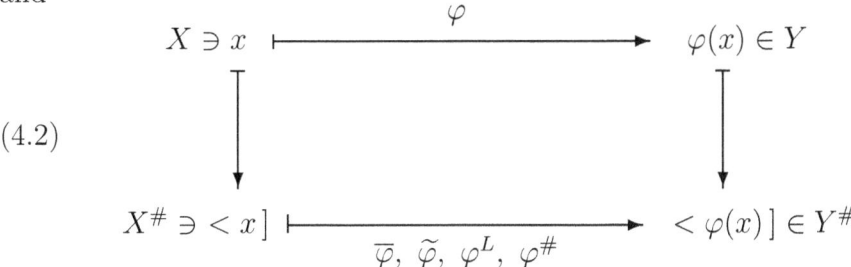

for every *cofinal* mapping L in (2.4).

Proof

It follows easily from the results in section 3.

Remark 4.1.

The extension in (4.1) does in fact *not* need the partial order on X, and it comes down to the extension in (A.28).

The extension in (4.2) comes down to the extension (A.29).

It follows that the extensions $\overline{\varphi}$, $\widetilde{\varphi}$ and φ^L, although not necessarily identical in general, do nevertheless reduce to $\varphi^{\#}$, in the case of the diagrams (4.1) and (4.2).

Appendix

We shortly present several notions and results used above. A related full presentation can be found in [3, Appendix, pp. 391-420].

Let $(X, \leq)$ be a nonvoid poset without minimum or maximum. For $a \in X$ we denote

(A.1) $\quad < a] = \{x \in X \mid x \leq a\}, \quad [a >= \{x \in X \mid x \geq a\}$

We define the mappings

(A.2) $\quad X \supseteq A \longmapsto A^u = \bigcap_{a \in A} [a > \subseteq X$

(A.3) $\quad X \supseteq A \longmapsto A^l = \bigcap_{a \in A} < a] \subseteq X$

then for $A \subseteq X$ we have

(A.4) $\quad A^u = X \iff A^l = X \iff A = \phi$

(A.5) $\quad A^u = \phi \iff A$ unbounded from above

(A.6) $\quad A^l = \phi \iff A$ unbounded from below

Definition A.1.

We call $A \subseteq X$ a *cut*, if and only if

(A.7) $\quad A^{ul} = A$

and denote

(A.8) $\quad X^{\#} = \{A \subseteq X \mid A \text{ is a cut}\} \subseteq \mathcal{P}(X)$

Clearly, (A.4) - (A.6) imply

(A.9) $\quad \phi, X \in X^{\#}$

therefore

(A.10) $\quad X^{\#} \neq \phi$

Given $A, B \subseteq X$, we have

(A.11) $\quad A \subseteq B \Longrightarrow A^u \supseteq B^u, \ A^l \supseteq B^l$

(A.12) $\quad A \subseteq A^{ul}, \quad A \subseteq A^{lu}$

(A.13) $\quad A^{ulu} = A^u, \quad A^{lul} = A^l$

Consequently

$$\forall \ A \subseteq X \ :$$

$$\ast) \ A^{ul} \in X^{\#}$$

(A.14) $\quad \ast\ast) \ \forall \ B \in X^{\#} \ :$

$$A \subseteq B \Longrightarrow A^{ul} \subseteq B$$

$$B \subseteq A \Longrightarrow B \subseteq A^{ul}$$

therefore

(A.15) $\quad X^{\#} = \{A^{ul} \mid A \subseteq X\}$

Given $x \in X$, we have

(A.16) $\quad \{x\}^u = [x >, \quad \{x\}^l = < x], \quad [x >^l = < x], \quad < x]^u = [x >$

143

(A.17) $\{x\}^{ul} = <x]$, $\{x\}^{lu} = [x>$

We denote for short

$\{x\}^u = x^u$, $\{x\}^l = x^l$, $\{x\}^{ul} = x^{ul}$, $\{x\}^{lu} = x^{lu}$, ...

Given $A \in X^{\#}$, we have

(A.18) $\phi \neq A \neq X \iff \begin{pmatrix} \exists \ a, b \in X : \\ <a] \subseteq A \subseteq <b] \end{pmatrix}$

We shall use the *embedding*

(A.19) $X \ni x \xmapsto{\varphi} x^{ul} = x^l = <x] \in X^{\#}$

We define on $X^{\#}$ the partial order

(A.20) $A \leq B \iff A \subseteq B$

Definition 2.1.

Given two posets $(X, \leq)$, $(Y, \leq)$ and a mapping $\varphi : X \longrightarrow Y$. We call φ an *order isomorphic embedding*, or in short, OIE, if and only if it is injective, and furthermore, for $a, b \in X$ we have

$$a \leq b \iff \varphi(a) \leq \varphi(b)$$

An OIE φ is an *order isomorphism*, or in short, OI, if and only if it is bijective, which in this case is equivalent with being surjective. □

The main result concerning order completion is given in, [2] :

Theorem (H M MacNeille, 1937)

1) The poset $(X^\#, \leq)$ is order complete.

2) The embedding $X \xrightarrow{\varphi} X^\#$ in (A.19) preserves infima and suprema, and it is an order isomorphic embedding, or OIE.

3) For $A \in X^\#$, we have the order density property of X in $X^\#$, namely

(A.21)
$$A = \sup\nolimits_{X^\#} \{x^l \mid x \in X, \ x^l \subseteq A\}$$
$$= \inf\nolimits_{X^\#} \{x^l \mid x \in X, \ A \subseteq x^l\}$$

$\square$

For $A \subseteq X$, we have

(A.22) $\quad A^{ul} = \sup\nolimits_{X^\#} \{x^l \mid x \in A\}$

Given $A_i \in X^\#$, with $i \in I$, we have with the partial order in $X^\#$ the relations

(A.23) $\quad \sup_{i \in I} A_i = \inf \{A \in X^\# \mid \bigcup_{i \in I} A_i \subseteq A\} = (\bigcup_{i \in I} A_i)^{ul}$

(A.24)
$\quad \inf_{i \in I} A_i = \sup \{A \in X^\# \mid A \subseteq \bigcap_{i \in I} A_i\} = (\bigcap_{i \in I} A_i)^{ul} =$
$\quad = \bigcap_{i \in I} A_i$

Extending mappings to order completions

Let $(X, \leq)$, $(Y, \leq)$ be two posets without minimum or maximum, and let

(A.25) $\quad \varphi : X \longrightarrow Y$

be any mapping. Our interest is to obtain an extension

$$\varphi^\# : X^\# \longrightarrow Y^\#$$

For that, we first extend φ to a *larger* domain, as follows

(A.26) $\quad \varphi^\# : \mathcal{P}(X) \longrightarrow Y^\#$

where for $A \subseteq X$ we define

(A.27) $\quad \varphi^\#(A) = (\varphi(A))^{ul} = \sup_{Y^\#} \{ <\varphi(x)] \mid x \in A \}$

and for any mapping in (A.25), we obtain the commutative diagram

(A.28)
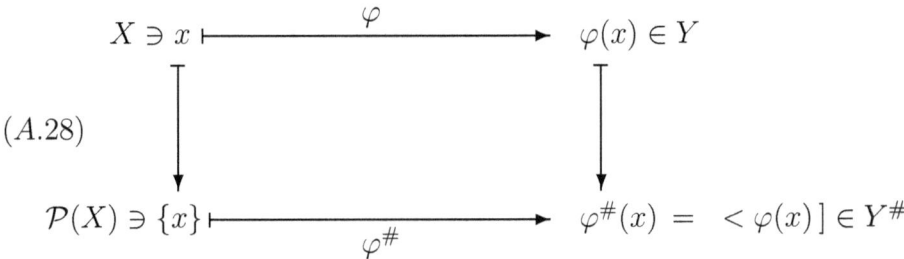

Proposition A.1.

1) The mapping $\varphi^\# : \mathcal{P}(X) \longrightarrow Y^\#$ in (A.36) is increasing, if on $\mathcal{P}(X)$ we take the partial order defined by the usual inclusion "$\subseteq$".

2) If the mapping $\varphi : X \longrightarrow Y$ in (A.35) is increasing, then the mapping $\varphi^\# : \mathcal{P}(X) \longrightarrow Y^\#$ in (A.36) is an extension of it to $X^\#$, namely, we have the commutative diagram

(A.29)
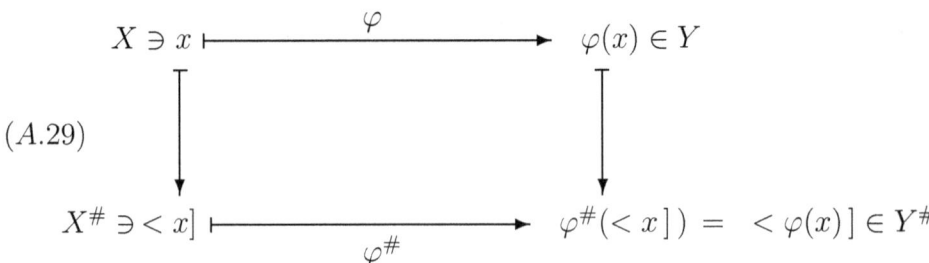

3) If the mapping $\varphi : X \longrightarrow Y$ in (A.25) is an OIE, then the mapping $\varphi^\# : \mathcal{P}(X) \longrightarrow Y^\#$ in (A.26) when restricted to $X^\#$, that is

(A.30) $\quad \varphi^\# : X^\# \longrightarrow Y^\#$

as in (A.29), is also an OIE.

Lemma A.1.

Let in general $\mu : M \longrightarrow N$ be an increasing mapping between two order complete posets, then for nonvoid $E \subseteq M$ we have

(A.31) $\quad \mu(\inf_M E) \leq \inf_N \mu(E) \leq \sup_N \mu(E) \leq \mu(\sup_M E)$

Proof

Indeed, let $a = \inf_M E \in M$. Then $a \leq b$, with $b \in E$. Hence $\mu(a) \leq \mu(b)$, with $b \in E$. Thus $\mu(a) \leq \inf_N \mu(E)$, and the first inequality is proved.
The last inequality is obtained in a similar manner, while the middle inequality is trivial.

$\square$

References

[1] Anguelov R, Rosinger E E : Hausdorff continuous solutions of nonlinear PDEs through the order completion method. Quaestiones Mathematicae, Vol. 28, 2005, 1-15, arXiv : math.AP/0406517

[2] Luxemburg W A J, Zaanen A C : Riesz Spaces, I. North-Holland, Amsterdam, 1971

[3] MacNeille H M : Partially ordered sets. Trans. AMS, Vol. 42, 1937, 416-460

[4] Oberguggenberger M B, Rosinger E E : Solution of Continuous Nonlinear PDEs through Order Completion. Mathematics Studies VOl. 181, North-Holland, Amsterdam, 1994

[5] Rosinger E E : Hausdorff continuous solutions of arbitrary continuous nonlinear PDEs through the order completion method. arXiv:math.AP/0405546

[6] Rosinger E E : Can there be a general nonlinear PDE theory for the existence of solutions ? arXiv:math.AP/0407026

[7] Rosinger E E : Solving large classes of nonlinear systems of PDEs. arXiv:math.AP/0505674

5. Solving Arbitrary Equations by Order Completion : Necessary and Sufficient Conditions for the Existence of Solutions

Solving General Equations by Order Completion, arXiv:math/0608450

Elemér E Rosinger

Department of Mathematics
and Applied Mathematics
University of Pretoria
Pretoria
0002 South Africa
eerosinger@hotmail.com

Abstract

A method based on *order completion* for solving general equations is presented. In particular, this method can be used for solving large classes of nonlinear systems of PDEs, with possibly associated initial and/or boundary value problems.

> "... provided also if need be that the notion of a solution shall be suitably extended ..."
>
> cited from Hilbert's 20th Problem

1. Preliminaries

Recently in [3], systems of nonlinear PDEs composed of equations of the general form

$$(1.1) \quad F(x, U(x), \ldots, D_x^p U(x), \ldots) = f(x), \quad x \in \Omega \subseteq \mathbb{R}^n$$

were solved on domains Ω that can be any open, not necessarily bounded subsets of $\mathbb{R}^n$, while $p \in \mathbb{N}^n$, $|p| \leq m$, with the orders $m \in \mathbb{N}$ of the PDEs arbitrary given.

The *unprecedented generality* of these nonlinear systems of PDEs comes, above all, from the class of functions F which define the left hand terms, and which are only assumed to be *jointly continuous* in all of their arguments. The right hand terms f are also required to be *continuous* only.

However, with minimal modifications of the method, both F and f can have certain *discontinuities* as well, [3].

Regardless of the above generality of the nonlinear systems of PDEs considered, and of possibly associated initial and/or boundary value problems, one can always find for them solutions U defined on the *whole* of the respective domains Ω. These solutions U have the *blanket, type independent*, or *universal regularity* property that they can be assimilated with *Hausdorff continuous functions*, [1,4-6].

It follows in this way that, when solving systems of nonlinear PDEs of the generality of those in (1.1), one can *dispense with* the various customary spaces of distributions, hyper-functions, generalized functions, Sobolev spaces, and so on. Instead one can stay within the realms of *usual functions*. Also, when proving the *existence* and the mentioned type of *regularity* of such solutions one can dispense with methods of Functional Analysis. However, functional analytic methods can possibly be used in order to obtain further regularity or other desirable properties of such solutions.

The mentioned generality of the equations solved and the regularity of the solutions obtained is based on the use of the *order completion method*, first introduced and developed in [3].
As it happens, however, this order completion method reaches far beyond the solution of systems of nonlinear PDEs, and in fact it can be applied to the solution of the surprisingly general equations

(1.2) $\quad T(A) = F$

where

(1.3) $\quad T : X \longrightarrow Y$

is any mapping, X is any nonvoid set, while $(Y, \leq)$ is a partially ordered set, or in short, poset, while $F \in Y$ is given, and $A \in X$ is the sought after solution.

Needless to say, in general, for a given $F \in Y$, there may not exist any solution $A \in X$ for the equation (1.2). Consequently, the setup (1.2), (1.3) may have to be *extended*.
Customarily, such extensions assume suitable topologies on X and Y, and certain continuity properties for the mapping T in (1.3).
However, as shown in [3], and also seen in the sequel, for the same purpose of solving the equations (1.2) in an extended setup, one can successfully use the *order completion* of the spaces X and Y.

And one of the major advantages of such an approach is that such a method

- does *no longer* differentiate between linear and nonlinear operators T in (1.3), in case the spaces X and Y may happen to have a linear vector space structure, [3].

Two particularly convenient further features of the order completion method in solving general equations (1.2) are the following :

- one obtains necessary and sufficient conditions for the existence of solutions,

- one obtains explicit expressions of the solutions, whenever they exist.

We shall present one of the general approaches resulting from the order completion method for solving equations of type (1.2). Further possible developments in this regard of the order completion method will be indicated.

2. Pull-Back Order

Without loss of generality, [3], we shall assume that all the posets considered are without a minimum or maximum element. Various notions and results related to partial orders which are used in the sequel are presented in the Appendix.

Given an equation (1.2), (1.3), we define on X the equivalence relation $\approx_T$ by

$$(2.1) \quad u \approx_T v \quad \Longleftrightarrow \quad T(u) = T(v)$$

for $u, v \in X$. In this way, by considering the quotient space

$$(2.2) \quad X_T = X/\approx_T$$

we obtain the *injective* mapping

$$(2.3) \quad T_\approx : X_T \longrightarrow Y$$

defined by

$$(2.4) \quad X_T \ni U \longmapsto T(u) \in Y$$

where $u \in U$, that is, U is the $\approx_T$ equivalence class of u in X_T, while $T(u)$ is defined by (1.3).

At that stage, we can define a partial order $\leq_T$ on X_T as being the *pull-back* by the mapping $T_\approx$ in (2.3) of the given partial order $\leq$ on Y, namely

(2.5) $\quad U \leq_T V \quad \Longleftrightarrow \quad T_\approx(U) \leq T_\approx(V)$

for $U, V \in X_T$. The effect of the above construction is that we obtain the *order isomorphic embedding*, or in short OIE

(2.6) $\quad T_\approx : X_T \longrightarrow Y$

As mentioned, without loss of generality we shall assume that the poset $(X_T^\#, \leq_T)$ has no minimum or maximum.

And now, we consider the *order completions* $X_T^\#$ and $Y^\#$ of X_T, and respectively, Y.
For simplicity, we shall denote by $\leq$ the partial orders both on $X_T^\#$ and $Y^\#$. In fact, as seen in (A.20), these partial orders are the usual inclusion relations $\subseteq$ among subsets of X, respectively, of Y.

Then according to Proposition A.1 in the Appendix, we obtain the commutative diagram of OIE-s

(2.7)
$$\begin{array}{ccc} X_T & \xrightarrow{T_\approx} & Y \\ \subseteq \downarrow & & \downarrow \subseteq \\ X_T^\# & \xrightarrow{T^\#} & Y^\# \end{array}$$

Consequently, for $U \in X_T$ and $A \in X_T^\#$, we have in $Y^\#$ the relations

(2.8) $\quad T^\#(<U]) \;=\; <T_\approx(U)]$

(2.9) $\quad T^{\#}(A) = (T_{\approx}(A))^{ul} = \sup_{Y^{\#}} \{ <T_{\approx}(U)] \mid U \in A \}$

3. Reformulation

Now we can reformulate the problem of solving the general equations (1.2) as follows. Given $F \in Y^{\#}$, find *necessary and sufficient* conditions for the existence of $A \in X_T^{\#}$, such that

(3.1) $\quad T^{\#}(A) = F$

4. Solution

We note that (2.7) gives the inclusions

(4.1)
$$\sup\nolimits_{Y^{\#}} \{ T^{\#}(U) \mid U \in X_T^{\#}, \, T^{\#}(U) \subseteq F \} \subseteq$$
$$\subseteq T^{\#}(\sup\nolimits_{X_T^{\#}} \{ U \mid U \in X_T^{\#}, \, T^{\#}(U) \subseteq F \}) \subseteq$$
$$\subseteq T^{\#}(\inf\nolimits_{X_T^{\#}} \{ V \mid V \in X_T^{\#}, \, F \subseteq T^{\#}(V) \}) \subseteq$$
$$\subseteq \inf\nolimits_{Y^{\#}} \{ T^{\#}(V) \mid V \in X_T^{\#}, \, F \subseteq T^{\#}(V) \}$$

Indeed, the first and last inclusions follow from Lemma A.1 in the Appendix. As for the middle inclusion in (4.1), let $U, V \in X_T^{\#}$ be such that $T^{\#}(U) \subseteq F \subseteq T^{\#}(V)$. Then $T^{\#}(U) \subseteq T^{\#}(V)$, hence $U \leq V$, since $T^{\#}$ is an OIE. It follows that

$$\sup\nolimits_{X_T^{\#}} \{ U \mid U \in X_T^{\#}, \, T^{\#}(U) \subseteq F \} \leq$$
$$\leq \inf\nolimits_{X_T^{\#}} \{ V \mid V \in X_T^{\#}, \, F \subseteq T^{\#}(V) \}$$

and the proof of (4.1) is completed.

We note further that the above inequality $T^{\#}(U) \subseteq F \subseteq T^{\#}(V)$ also implies

$$\sup\nolimits_{Y^{\#}} \{ T^{\#}(U) \mid U \in X_T^{\#}, \ T^{\#}(U) \subseteq F \} \subseteq F \subseteq$$
(4.2)
$$\subseteq \inf\nolimits_{Y^{\#}} \{ T^{\#}(V) \mid V \in X_T^{\#}, \ F \subseteq T^{\#}(V) \}$$

Furthermore

(4.3) $\quad \{ U \in X_T^{\#} \mid T^{\#}(U) \subseteq F \} \neq \phi$

since (A.9) gives $U = \phi \in X_T^{\#}$, hence in view of (A.4) - (A.6) and (2.9) we have $T^{\#}(U) = (T(\phi))^{ul} = \phi^{ul} = (\phi^u)^l = (Y^{\#})^l = \phi \subseteq F$.

Returning now to the problem (3.1), we note that it is not trivial. Indeed, the OIE in (2.7), namely

$$T^{\#} : X_T^{\#} \longrightarrow Y^{\#}$$

need *not* be surjective. Further $T^{\#}$ need *not* preserve infima or suprema.

However, in the next theorem we can obtain the following two general results :

- a necessary and sufficient condition for the solvability of (9,12), and

- the explicit expression of the solution, when it exists.

Theorem 4.1.

Given $F \in Y^{\#}$.

1) The equation

(4.4) $\quad T^{\#}(A) = F$

has a solution $A \in X_T^{\#}$, if and only if, see (4.1)

(4.5) $\sup_{Y^\#} \{ T^\#(U) \mid U \in X_T^\#, T^\#(U) \subseteq F \} =$
$$= \inf_{Y^\#} \{ T^\#(V) \mid V \in X_T^\#, F \subseteq T^\#(V) \}$$

2) This solution is unique, whenever it exists, see (2.7).

3) When it exists, the unique solution $A \in X_T^\#$ is given by

(4.6) $A = \sup_{X_T^\#} \{ U \in X_T^\# \mid T^\#(U) \subseteq F \} =$
$$= \inf_{X_T^\#} \{ V \in X_T^\# \mid F \subseteq T^\#(V) \}$$

and, see (4.3)

(4.7) $\{ U \in X_T^\# \mid T^\#(U) \subseteq F \}, \{ V \in X_T^\# \mid F \subseteq T^\#(V) \} \neq \phi$

Proof

From (4.4) follows that $A \in X_T^\#$, $T^\#(A) \subseteq F$, thus

$$F = T^\#(A) \subseteq \sup_{Y^\#} \{ T^\#(U) \mid U \in X_T^\#, T^\#(U) \subseteq F \}$$

Similarly we have

$$\inf_{Y^\#} \{ T^\#(V) \mid V \in X_T^\#, F \subseteq T^\#(V) \} \subseteq T^\#(A) = F$$

thus (4.1) collapses to the seven equalities

$$F = T^\#(A) = \sup_{Y^\#} \ldots = T^\#(\sup_{X_T^\#} \ldots) =$$
$$= T^\#(\inf_{X_T^\#} \ldots) = \inf_{Y^\#} \ldots = T^\#(A) = F$$

Thus in particular we obtain (4.5).

The injectivity of $T^\#$ will give (4.6), while (4.7) follows from (4.3) and the fact that we can take $V = A$.

Conversely, let us assume (4.5). Then (4.1) collapses to the three equalities

$$\sup\nolimits_{Y^\#} \ldots = T^\#(\sup\nolimits_{X_T^\#} \ldots) = T^\#(\inf\nolimits_{X_T^\#} \ldots) = \inf\nolimits_{Y^\#} \ldots$$

thus in view of the corresponding collapsed version of (4.2), we can extend the above three equalities to the following four

$$\sup\nolimits_{Y^\#} \ldots = T^\#(\sup\nolimits_{X_T^\#} \ldots) = T^\#(\inf\nolimits_{X_T^\#} \ldots) = \inf\nolimits_{Y^\#} \ldots = F$$

And now the injectivity of $T^\#$ will give (4.4) and (4.6), while (4.7) follows as above.

□

The above existence result is of a "local" nature, since it refers to a solution of the equation (3.1) for one given right hand term $F \in Y^\#$. This result, however, can further be strengthened by the following "global" one which characterizes the solvability of (3.1) for *all* right hand terms $F \in Y^\#$. Namely, we have, [3, pp. 190,191]

Theorem 4.2.

The following are equivalent

(4.8) $T^\#(X_T^\#) \supseteq Y$

and

(4.9) $T^\#(X_T^\#) = Y^\#$

In each of these cases $T^\#$ is an order isomorphism, or in short, OI, between $X_T^\#$ and $Y^\#$.

It is *important* to note the following two fact :

- "Pull-back" type structures are customary when solving PDEs by functional analytic methods. Details in this regard are presented in [3, chap. 12], while one well known classical example can be seen in section 7 in the sequel.

- As shown in [3, chap. 13], one can consider in (2.7) far more general partial orders than the "pull-back" type ones, and still obtain solutions for nonlinear PDEs in (1.1) by the order completion method.

5. Applications to Nonlinear Systems of PDEs

Let us now associate with a nonlinear PDE in (1.1) the corresponding nonlinear partial differential operator defined by the left hand side, namely

(5.1) $\quad T(x,D)U(x) = F(x, U(x), \ldots, D_x^p U(x), \ldots), \quad x \in \Omega$

Two facts about the nonlinear PDEs in (1.1) and the corresponding nonlinear partial differential operators $T(x,D)$ in (5.1) are important and immediate :

- The operators $T(x,D)$ can *naturally* be seen as acting in the *classical* context, namely

(5.2) $\quad T(x,D) : \mathcal{C}^m(\Omega) \ni U \longmapsto T(x,D)U \in \mathcal{C}^0(\Omega)$

while, unfortunately on the other hand :

- The mappings in this natural classical context (5.2) are typically *not* surjective even in the case of linear $T(x,D)$, and they are even less so in the general nonlinear case of (1.1).

In other words, linear or nonlinear PDEs in (1.1) typically *cannot* be expected to have *classical* solutions $U \in \mathcal{C}^m(\Omega)$, for arbitrary continuous right hand terms $f \in \mathcal{C}^0(\Omega)$, as illustrated by a variety of well known examples, some of them rather simple ones, see [3, chap. 6]. Furthermore, it can often happen that non-classical solutions do have a major applicative interest, thus they have to be sought out *beyond* the confines of the classical framework in (5.2).
This is, therefore, how we are led to the *necessity* to consider *generalized solutions* U for PDEs like those in (1.1), that is, solutions

$U \notin \mathcal{C}^m(\Omega)$, which therefore are no longer classical. This means that the natural classical mappings (5.2) must in certain suitable ways be *extended* to *commutative diagrams*

(5.3)
$$\begin{array}{ccc} \mathcal{C}^m(\Omega) & \xrightarrow{T(x,D)} & \mathcal{C}^0(\Omega) \\ \cap\big\downarrow & & \cap\big\downarrow \\ X & \xrightarrow{\overline{T}} & Y \end{array}$$

with the generalized solutions now being found as

(5.4) $\quad U \in X \setminus \mathcal{C}^m(\Omega)$

instead of the classical ones $U \in \mathcal{C}^m(\Omega)$ which may easily fail to exist. A further important point is that one expects to reestablish certain kind of *surjectivity* type properties typically missing in (5.2), at least such as for instance

(5.5) $\quad \mathcal{C}^0(\Omega) \subseteq \overline{T}(X)$

As it turns out, when constructing extensions of (5.2) given by commutative diagrams (5.3), we shall be interested in the following somewhat larger spaces of piecewise smooth functions. For any integer $0 \leq l \leq \infty$, we define

(5.6) $\quad \mathcal{C}^l_{nd}(\Omega) = \left\{ u : \Omega \to \mathbb{R} \;\middle|\; \begin{array}{l} \exists\, \Gamma \subset \Omega \text{ closed, nowhere dense :} \\ u \in \mathcal{C}^l(\Omega \setminus \Gamma) \end{array} \right\}$

and as an immediate strengthening of (5.2), we obviously obtain

(5.7) $\quad T(x,D)\, \mathcal{C}^m_{nd}(\Omega) \subseteq \mathcal{C}^0_{nd}(\Omega)$

The solution of the nonlinear PDEs in (1.1) through the order completion method will come from the construction of specific instances of the *commutative diagrams* (5.3), given by

(5.8)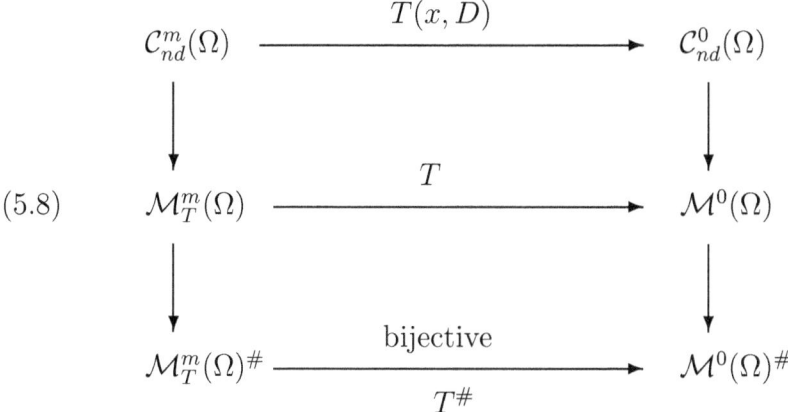

where the operation $(\ )^{\#}$ means the *order completion*, [3], of the respective spaces, as well as the extension to such order completions of the respective mappings, see (2.7). It follows that in terms of (5.3), we have

$$X = \mathcal{M}_T^m(\Omega)^{\#}, \quad Y = \mathcal{M}^0(\Omega)^{\#}, \quad \overline{T} = T^{\#}$$

thus we shall obtain for the nonlinear PDEs in (1.1) generalized solutions

(5.9) $U \in \mathcal{M}_T^m(\Omega)^{\#}$

Furthermore, instead of the *surjectivity* condition (5.5), we shall at least have the following stronger one

(5.10) $\mathcal{C}_{nd}^0(\Omega) \subseteq T^{\#}(\mathcal{M}_T^m(\Omega)^{\#})$

So far about the main ideas related to the *existence* of solutions of general nonlinear PDEs of the form (1.1). Further details can be found in [3,1,4-6].

As for the *regularity* of such solutions, we recall that, as shown in [1], one has the inclusions

(5.11) $\mathcal{M}^0(\Omega)^{\#} \subseteq Mes\,(\Omega)$

where $Mes\,(\Omega)$ denotes the set of Lebesgue measurable functions on Ω. In this way, in view of (5.8) and (5.9), one can assimilate the generalized solutions U of the nonlinear PDEs in (1.1) with usual measurable functions in $Mes\,(\Omega)$.

Recently, however, based on results in [1,4-6], it was shown that instead of (5.11), one has the much *stronger regularity* property

$$(5.12) \quad \mathcal{M}^0(\Omega)^{\#} \subseteq \mathbb{H}\,(\Omega)$$

where $\mathbb{H}\,(\Omega)$ denotes the set of Hausdorff continuous functions on Ω. Consequently, now one can significantly improve on the earlier regularity result, as one can assimilate the generalized solutions U of the nonlinear PDEs in (1.1) with usual functions in $\mathbb{H}\,(\Omega)$.

Regarding *systems* of nonlinear PDEs such as in (1.1), with possibly associated initial and/or boundary value problems, it was shown in [3, chap. 8] the way they can be dealt with the above order completion method.

In this respect, a *surprising* advantage of the order completion method is the ease, when compared with the usual functional analytic approaches, in dealing with initial and/or boundary value problems.

6. Beyond "Pull-Back" Partial Orders

As mentioned at the end of section 4, and presented in full detail in [3, chap. 13], the order completion method in solving large classes of nonlinear systems of PDEs of the type (1.1) is *not* limited to the use of "pull-back" type partial orders in (2.7), (5.3) and (5.8). In fact, a large class of more general partial orders can be defined on the domains $\mathcal{M}_T^m(\Omega)$ of the respective PDEs, and stil obtain for them solutions in the corresponding order completions.

7. Use of "Pull-Back" in Functional Analytic Solution Methods

As presented in detail in [3, chap. 12], functional analytic methods used for solving PDEs do often employ topologies obtained by "pull-back". Here we present shortly one of the classical such examples. Let us consider on a bounded Euclidean domain Ω, which has a smooth boundary $\partial \Omega$, the following familiar linear boundary value problem, usually called the Poisson Problem

(7.1)
$$\Delta U(x) = f(x), \quad x \in \Omega$$
$$U = 0 \quad \text{on } \partial \Omega$$

As is well known, for every given $f \in C^\infty(\overline{\Omega})$, where $\overline{\Omega}$ denotes the closure of Ω, this problem has a unique solution U in the space

(7.2) $\quad X = \{\, v \in C^\infty(\overline{\Omega}) \mid v = 0 \text{ on } \partial \Omega \,\}$

It follows that the mapping

(7.3) $\quad X \ni v \longmapsto \|\Delta v\|_{L^2(\Omega)}$

defines a norm on the vector space X. Now let

(7.4) $\quad Y = C^\infty(\overline{\Omega})$

be endowed with the topology induced by $L^2(\Omega)$. Then in view of (7.1) - (7.4), the mapping

(7.5) $\quad \Delta : X \to Y$

is a uniform continuous linear bijection. Therefore, it can be extended in a unique manner to an isomorphism of Banach spaces

(7.6) $\quad \Delta : \overline{X} \to \overline{Y} = L^2(\Omega)$

In this way one has the classical existence and uniqueness result

$$\forall\ f \in L^2(\Omega)\ :$$

(7.7) $\quad \exists !\ U \in \overline{X}\ :$

$$\Delta U = f$$

The power and simplicity - based on linearity and topological completion of uniform spaces - of the above classical existence and uniqueness result is obvious. This power is illustrated by the fact that the set $\overline{Y} = L^2(\Omega)$ in which the right hand terms f in (7.1) can now be chosen is much *larger* than the original $Y = C^\infty(\overline{\Omega})$. Furthermore, the existence and uniqueness result in (7.7) does not need the a priori knowledge of the structure of the elements $U \in \overline{X}$, that is, of the respective generalized solutions. This structure which gives the regularity properties of such solutions can be obtained by a further detailed study of the respective differential operators defining the PDEs under consideration, in this case, the Laplacean Δ. And in the above specific instance we obtain

(7.8) $\quad \overline{X} = H^2(\Omega) \cap H^1_0(\Omega)$

As seen above, typically for the functional analytic methods, the generalized solutions are obtained in topological completions of vector spaces of usual functions. And such completions, like for instance the various Sobolev spaces, are defined by certain linear partial differential operators which may happen to *depend* on the PDEs under consideration.

In the above example, for instance, the topology on the space X obviously *depends* on the specific PDE in (7.1). Thus the topological completion $\overline{X}$ in which the generalized solutions U are found according to (7.7), does again *depend* on the respective PDE.

Appendix

We shortly present several notions and results used above. A related full presentation can be found in [3, Appendix, pp. 391-420].

Let $(X, \leq)$ be a nonvoid poset without minimum or maximum. For $a \in X$ we denote

(A.1) $\quad <a] = \{x \in X \mid x \leq a\}, \quad [a> = \{x \in X \mid x \geq a\}$

We define the mappings

(A.2) $\quad X \supseteq A \longmapsto A^u = \bigcap_{a \in A} [a> \subseteq X$

(A.3) $\quad X \supseteq A \longmapsto A^l = \bigcap_{a \in A} <a] \subseteq X$

then for $A \subseteq X$ we have

(A.4) $\quad A^u = X \iff A^l = X \iff A = \phi$

(A.5) $\quad A^u = \phi \iff A$ unbounded from above

(A.6) $\quad A^l = \phi \iff A$ unbounded from below

Definition A.1.

We call $A \subseteq X$ a *cut*, if and only if

(A.7) $\quad A^{ul} = A$

and denote

(A.8) $\quad X^{\#} = \{A \subseteq X \mid A \text{ is a cut}\} \subseteq \mathcal{P}(X)$

$\square$

Clearly, (A.4) - (A.6) imply

(A.9) $\quad \phi, X \in X^{\#}$

therefore

163

(A.10) $\quad X^\# \neq \phi$

Given $A, B \subseteq X$, we have

(A.11) $\quad A \subseteq B \Longrightarrow A^u \supseteq B^u, \ A^l \supseteq B^l$

(A.12) $\quad A \subseteq A^{ul}, \quad A \subseteq A^{lu}$

(A.13) $\quad A^{ulu} = A^u, \quad A^{lul} = A^l$

Consequently

$$\forall \ A \subseteq X :$$

$$*) \ A^{ul} \in X^\#$$

(A.14) $\quad **) \ \forall \ B \in X^\# :$

$$A \subseteq B \Longrightarrow A^{ul} \subseteq B$$

$$B \subseteq A \Longrightarrow B \subseteq A^{ul}$$

therefore

(A.15) $\quad X^\# = \{A^{ul} \mid A \subseteq X\}$

Given $x \in X$, we have

(A.16) $\quad \{x\}^u = [x >, \quad \{x\}^l =< x], \quad [x >^l =< x], \quad < x]^u = [x >$

(A.17) $\quad \{x\}^{ul} =< x], \quad \{x\}^{lu} = [x >$

We denote for short

$$\{x\}^u = x^u, \quad \{x\}^l = x^l, \quad \{x\}^{ul} = x^{ul}, \quad \{x\}^{lu} = x^{lu}, \ \ldots$$

Given $A \in X^\#$, we have

$$(A.18) \quad \phi \neq A \neq X \iff \begin{pmatrix} \exists\ a, b \in X\ : \\ <a] \subseteq A \subseteq <b] \end{pmatrix}$$

We shall use the *embedding*

$$(A.19) \quad X \ni x \xmapsto{\varphi} x^{ul} = x^l = <x] \in X^{\#}$$

We define on $X^{\#}$ the partial order

$$(A.20) \quad A \leq B \iff A \subseteq B$$

Definition 2.1.

Given two posets $(X, \leq)$, $(Y, \leq)$ and a mapping $\varphi : X \longrightarrow Y$. We call φ an *order isomorphic embedding*, or in short, OIE, if and only if it is injective, and furthermore, for $a, b \in X$ we have

$$a \leq b \iff \varphi(a) \leq \varphi(b)$$

An OIE φ is an *order isomorphism*, or in short, OI, if and only if it is bijective.

□

The main result concerning order completion is given in, [2] :

Theorem (H M MacNeille, 1937)

1) The poset $(X^{\#}, \leq)$ is order complete.

2) The embedding $X \xrightarrow{\varphi} X^{\#}$ in (A.19) preserves infima and suprema, and it is an order isomorphic embedding, or OIE.

3) For $A \in X^{\#}$, we have the order density property of X in $X^{\#}$, namely

(A.21)
$$A = \sup_{X^\#} \{x^l \mid x \in X, \; x^l \subseteq A\} =$$
$$= \inf_{X^\#} \{x^l \mid x \in X, \; A \subseteq x^l\}$$

$\square$

For $A \subseteq X$, we have

(A.22) $\quad A^{ul} = \sup_{X^\#} \{x^l \mid x \in A\}$

Given $A_i \in X^\#$, with $i \in I$, we have with the partial order in $X^\#$ the relations

(A.23) $\quad \sup_{i \in I} A_i = \inf \{A \in X^\# \mid \bigcup_{i \in I} A_i \subseteq A\} = (\bigcup_{i \in I} A_i)^{ul}$

(A.24)
$$\inf_{i \in I} A_i = \sup \{A \in X^\# \mid A \subseteq \bigcap_{i \in I} A_i\} = (\bigcap_{i \in I} A_i)^{ul} =$$
$$= \bigcap_{i \in I} A_i$$

Extending mappings to order completions

Let $(X, \leq)$, $(Y, \leq)$ be two posets without minimum or maximum, and let

(A.25) $\quad \varphi : X \longrightarrow Y$

be any mapping. Our interest is to obtain an extension

$$\varphi^\# : X^\# \longrightarrow Y^\#$$

For that, we first extend φ to a *larger* domain, as follows

(A.26) $\quad \varphi^\# : \mathcal{P}(X) \longrightarrow Y^\#$

where for $A \subseteq X$ we define

(A.27) $\quad \varphi^\#(A) = (\varphi(A))^{ul} = \sup_{Y^\#} \{< \varphi(x)] \mid x \in A\}$

and for any mapping in (A.25), we obtain the commutative diagram

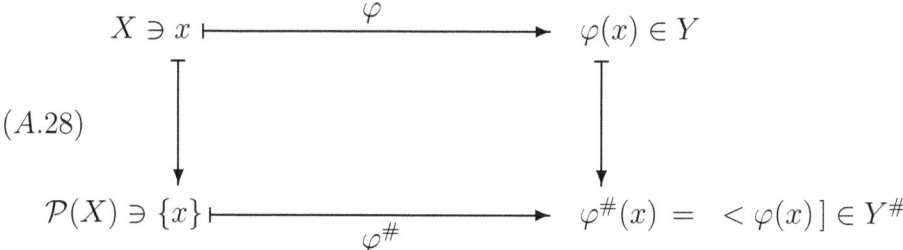

(A.28)

Proposition A.1.

1) The mapping $\varphi^{\#} : \mathcal{P}(X) \longrightarrow Y^{\#}$ in (A.36) is increasing, if on $\mathcal{P}(X)$ we take the partial order defined by the usual inclusion "$\subseteq$".

2) If the mapping $\varphi : X \longrightarrow Y$ in (A.35) is increasing, then the mapping $\varphi^{\#} : \mathcal{P}(X) \longrightarrow Y^{\#}$ in (A.36) is an extension of it to $X^{\#}$, namely, we have the commutative diagram

$$
\begin{array}{ccc}
X \ni x & \xrightarrow{\varphi} & \varphi(x) \in Y \\
\downarrow & & \downarrow \\
X^{\#} \ni <x] & \xrightarrow{\varphi^{\#}} & \varphi^{\#}(<x]) = <\varphi(x)] \in Y^{\#}
\end{array}
$$

(A.29)

3) If the mapping $\varphi : X \longrightarrow Y$ in (A.25) is an OIE, then the mapping $\varphi^{\#} : \mathcal{P}(X) \longrightarrow Y^{\#}$ in (A.26) when restricted to $X^{\#}$, that is

(A.30) $\varphi^{\#} : X^{\#} \longrightarrow Y^{\#}$

as in (A.29), is also an OIE.

Lemma A.1.

Let in general $\mu : M \longrightarrow N$ be an increasing mapping between two order complete posets, then for nonvoid $E \subseteq M$ we have

(A.31) $\quad \mu(\inf_M E) \leq \inf_N \mu(E) \leq \sup_N \mu(E) \leq \mu(\sup_M E)$

Proof

Indeed, let $a = \inf_M E \in M$. Then $a \leq b$, with $b \in E$. Hence $\mu(a) \leq \mu(b)$, with $b \in E$. Thus $\mu(a) \leq \inf_N \mu(E)$, and the first inequality is proved.
The last inequality is obtained in a similar manner, while the middle inequality is trivial.

$\square$

References

[1] Anguelov R, Rosinger E E : Hausdorff continuous solutions of nonlinear PDEs through the order completion method. Quaestiones Mathematicae, Vol. 28, 2005, 1-15, arXiv : math.AP/0406517

[2] Luxemburg W A J, Zaanen A C : Riesz Spaces, I. North-Holland, Amsterdam, 1971

[3] Oberguggenberger M B, Rosinger E E : Solution of Continuous Nonlinear PDEs through Order Completion. Mathematics Studies VOl. 181, North-Holland, Amsterdam, 1994

[4] Rosinger E E : Hausdorff continuous solutions of arbitrary continuous nonlinear PDEs through the order completion method. math.AP/0405546

[5] Rosinger E E : Can there be a general nonlinear PDE theory for the existence of solutions ? math.AP/0407026

[6] Rosinger E E : Solving large classes of nonlinear systems of PDEs. math.AP/0505674

6. Further Details on Solving PDEs by Order Completion

New Method for Solving Large Classes of Nonlinear Systems of PDEs

Elemér E Rosinger

*Department of Mathematics
and Applied Mathematics
University of Pretoria
Pretoria
0002 South Africa
eerosinger@hotmail.com*

Abstract

The essentials of a new method in solving very large classes of nonlinear systems of PDEs, possibly associated with initial and/or boundary value problems, are presented. The PDEs can be defined by continuous, not necessarily smooth expressions, and the solutions obtained cab be assimilated with usual measurable functions, or even with Hausdorff continuous ones. The respective result sets aside completely, and with a large nonlinear margin, the celebrated 1957 impossibility of Hans Lewy regarding the nonexistence of solution in distributions of large classes of linear smooth coefficient PDEs.

1. The Class of Nonlinear Systems of PDEs Solved

The nonlinear systems of PDEs considered in this paper are composed of equations of the general form

(1.1) $\quad F(x, U(x), \ldots, D_x^p U(x), \ldots) = f(x), \quad x \in \Omega \subseteq \mathbb{R}^n$

were the domains Ω can be any open, not necessarily bounded subsets of $\mathbb{R}^n$, while $p \in \mathbb{N}^n$, $|p| \leq m$, with the orders $m \in \mathbb{N}$ of the PDEs arbitrary given.

The functions F which define the left hand terms are only assumed to be *jointly continuous* in all of their arguments. The right hand terms f are also required to be *continuous* only.
However, with minimal modifications of the method, both F and f can have certain *discontinuities* as well, [3].

As it turns out, regardless of the above generality of the nonlinear systems of PDEs considered, and of possibly associated initial and/or boundary value problems, one can always find for them solutions U defined on the *whole* of the respective domains Ω. These solutions U have the *blanket, type independent*, or *universal regularity* property that they can be assimilated with usual *measurable*, or even *Hausdorff continuous functions*, [1,4-6].

Thus when solving systems of nonlinear PDEs of the generality of those in (1.1), one can *dispense with* the various customary spaces of distributions, hyper-functions, generalized functions, Sobolev spaces, and so on. Instead one can stay within the realms of *usual functions*. However, functional analytic methods can possibly be used in order to obtain further regularity or other desirable properties of such solutions.

2. A Natural Smooth Framework for Nonlinear Systems of PDEs

Let us now associate with a nonlinear PDE in (1.1) the corresponding nonlinear partial differential operator defined by the left hand side, namely

(2.1) $\quad T(x, D)U(x) = F(x, U(x), \ldots, D_x^p U(x), \ldots), \quad x \in \Omega$

Two facts about the nonlinear PDEs in (1.1) and the corresponding nonlinear partial differential operators $T(x, D)$ in (2.1) are important and immediate :

- The operators $T(x, D)$ can *naturally* be seen as acting in the *classical* or smooth context, namely

(2.2) $\quad T(x, D) : \mathcal{C}^m(\Omega) \ni U \longmapsto T(x, D)U \in \mathcal{C}^0(\Omega)$

while, unfortunately on the other hand :

- The mappings in this natural classical context (2.2) are typically *not* surjective even in the case of linear $T(x, D)$, and they are even less so in the general nonlinear case of (1.1).

In other words, linear or nonlinear PDEs in (1.1) typically *cannot* be expected to have *classical* solutions $U \in \mathcal{C}^m(\Omega)$, for arbitrary continuous right hand terms $f \in \mathcal{C}^0(\Omega)$, as illustrated by a variety of well known examples, some of them rather simple ones, see [3, chap. 6].
Furthermore, it can often happen that non-classical solutions do have a major applicative interest, thus they have to be sought out *beyond* the confines of the classical framework in (2.2).
This is, therefore, how we are led to the *necessity* to consider *generalized solutions* U for PDEs like those in (1.1), that is, solutions $U \notin \mathcal{C}^m(\Omega)$, which therefore are no longer classical. This means that the natural classical mappings (2.2) must in certain suitable ways be *extended* to commutative diagrams

(2.3)

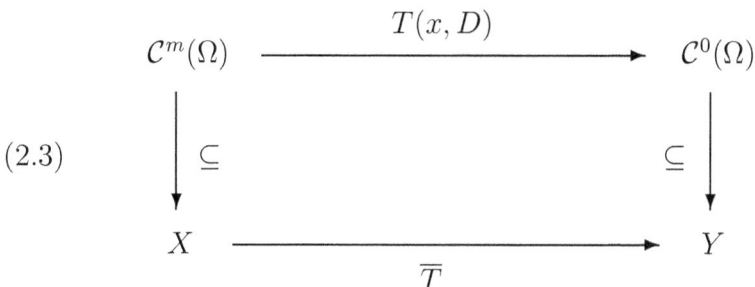

with the generalized solutions now being found as

$$(2.4) \quad U \in X \setminus \mathcal{C}^m(\Omega)$$

instead of the classical ones $U \in \mathcal{C}^m(\Omega)$ which may easily fail to exist. A further important point is that one expects to reestablish certain kind of *surjectivity* type properties typically missing in (2.2), at least such as for instance

$$(2.5) \quad \mathcal{C}^0(\Omega) \subseteq \overline{T}(X)$$

As it turns out, when constructing extensions of (2.2) given by commutative diagrams (2.3), we shall be interested in the following somewhat larger spaces of piecewise smooth functions. For any integer $0 \leq l \leq \infty$, we define

$$(2.6) \quad \mathcal{C}^l_{nd}(\Omega) = \left\{ u : \Omega \to \mathbb{R} \;\middle|\; \begin{array}{l} \exists \, \Gamma \subset \Omega \text{ closed, nowhere dense :} \\ u \in \mathcal{C}^l(\Omega \setminus \Gamma) \end{array} \right\}$$

and as an immediate strengthening of (2.2), we obviously obtain

$$(2.7) \quad T(x, D) \, \mathcal{C}^m_{nd}(\Omega) \subseteq \mathcal{C}^0_{nd}(\Omega)$$

The solution of the nonlinear PDEs in (1.1) through the order completion method will come from the construction of specific instances of the *commutative diagrams* (2.3), given by

$$(2.8) \quad \begin{array}{ccc} \mathcal{C}^m_{nd}(\Omega) & \xrightarrow{T(x,D)} & \mathcal{C}^0_{nd}(\Omega) \\ \downarrow & & \downarrow \\ \mathcal{M}^m_T(\Omega) & \xrightarrow{T} & \mathcal{M}^0(\Omega) \\ \downarrow & & \downarrow \\ \mathcal{M}^m_T(\Omega)^\# & \xrightarrow[T^\#]{\text{bijective}} & \mathcal{M}^0(\Omega)^\# \end{array}$$

where the operation ( )# means the *order completion*, [3], of the respective spaces, as well as the extension to such order completions of the respective mappings, see (5.7) below. It follows that in terms of (2.3), we have

$$X = \mathcal{M}_T^m(\Omega)^\#, \quad Y = \mathcal{M}^0(\Omega)^\#, \quad \overline{T} = T^\#$$

thus we shall obtain for the nonlinear PDEs in (1.1) generalized solutions

(2.9) $\quad U \in \mathcal{M}_T^m(\Omega)^\#$

Furthermore, instead of the *surjectivity* condition (2.5), we shall at least have the following stronger one

(2.10) $\quad \mathcal{C}_{nd}^0(\Omega) \subseteq T^\#(\mathcal{M}_T^m(\Omega)^\#)$

So far about the main ideas related to the *existence* of solutions of general nonlinear PDEs of the form (1.1). Further details can be found in [3,1,4-6].

As for the *regularity* of such solutions, we recall that, as shown in [1], one has the inclusions

(2.11) $\quad \mathcal{M}^0(\Omega)^\# \subseteq Mes(\Omega)$

where $Mes(\Omega)$ denotes the set of Lebesgue measurable functions on Ω. In this way, in view of (2.8) and (2.9), one can assimilate the generalized solutions U of the nonlinear PDEs in (1.1) with usual measurable functions in $Mes(\Omega)$.
Recently, however, based on results in [1,4-6], it was shown that instead of (2.11), one has the much *stronger regularity* property

(2.12) $\quad \mathcal{M}^0(\Omega)^\# \subseteq \mathbb{H}(\Omega)$

where $\mathbb{H}(\Omega)$ denotes the set of Hausdorff continuous functions on Ω. Consequently, now one can significantly improve on the earlier regularity result, as one can assimilate the generalized solutions U of the nonlinear PDEs in (1.1) with usual functions in $\mathbb{H}(\Omega)$.

Regarding *systems* of nonlinear PDEs such as in (1.1), with possibly associated initial and/or boundary value problems, it was shown in [3, chap. 8] the way they can be dealt with the above order completion method.

In this respect, a *surprising* advantage of the order completion method is the ease, when compared with the usual functional analytic approaches, in dealing with initial and/or boundary value problems.

3. Solution Method by Order Completion

As first introduced and developed in [3], the mentioned *existence* and *blanket, type independent,* or *universal regularity* property of solutions for nonlinear systems of PDEs of the above generality in (1.1), together with possibly associated initial and/or boundary value problems, is based on a rather simple, however, quite general *order completion* of suitable spaces of piecewise smooth functions.

Here one can mention that, in fact, this order completion based solution method reaches *far beyond* the solution of PDEs, and in fact it can be applied to the solution of general equations

$$(3.1) \quad T(A) = F$$

where

$$(3.2) \quad T : X \longrightarrow Y$$

is any mapping, X is any nonvoid set, while $(Y, \leq)$ is a partially ordered set, or in short, poset, while $F \in Y$ is given, and $A \in X$ is the sought after solution.

In general, for a given $F \in Y$, there may not exist any solution $A \in X$ for the equation (3.2), unless the setup (3.1), (3.2) is *extended* in suitable ways.

The usual such extensions assume suitable topologies on X and Y, and certain continuity properties for the mapping T in (3.2).

As shown in [3], and also seen in the sequel, for the same purpose of solving the equations (3.1) in a suitably extended setup, one can successfully use the *order completion* of the spaces X and Y.

A *major advantages* of the order completion approach is that such a method

- does *no longer* differentiate between linear and nonlinear operators T in (3.2), in case the spaces X and Y may happen to have a linear vector space structure, [3].

Two further convenient features of the order completion method in solving general equations (3.1) are the following :

- one obtains necessary and sufficient conditions for the existence of solutions,

- one obtains explicit expressions of the solutions, whenever they exist.

The order completion method for solving general equations (3.1) presented here is but one of the possible such approaches. This method is based on so called "pull-back" partial orders on the domains of the mappings T in (3.1), (3.2). However, the order completion method for solving general equations (3.1) is *not* limited to "pull-back" partial orders. Further possible developments in this regard of the order completion method, and *no longer* based on "pull-back" partial orders will be indicated.

Here it should be mentioned that usual functional analytic methods for solving linear or nonlinear PDEs often make use of "pull-back" topologies on the domains of the respective partial differential operators as illustrated next in section 4, see for further details in [3, chap. 12].

4. Use of "Pull-Back" in Functional Analytic Solution Methods of PDEs

We present here one of the classical examples where a "pull-back" topology on the domain of a partial differential operator is used in order to solve the corresponding PDE. Let us consider on a bounded Euclidean domain Ω, which has a smooth boundary $\partial \Omega$, the following familiar linear boundary value problem, usually called the Poisson Problem

(4.1)
$$\Delta U(x) = f(x), \quad x \in \Omega$$
$$U = 0 \quad \text{on} \quad \partial \Omega$$

As is well known, for every given $f \in C^\infty(\overline{\Omega})$, where $\overline{\Omega}$ denotes the closure of Ω, this problem has a unique solution U in the space

(4.2) $\quad X = \{\, v \in C^\infty(\overline{\Omega}) \mid v = 0 \text{ on } \partial \Omega \,\}$

It follows that the mapping

(4.3) $\quad X \ni v \longmapsto \|\Delta v\|_{L^2(\Omega)}$

defines a norm on the vector space X. Now let

(4.4) $\quad Y = C^\infty(\overline{\Omega})$

be endowed with the topology induced by $L^2(\Omega)$. Then in view of (4.1) - (4.4), the mapping

(4.5) $\quad \Delta : X \to Y$

is a uniform continuous linear bijection. Therefore, it can be extended in a unique manner to an isomorphism of Banach spaces

(4.6) $\quad \Delta : \overline{X} \to \overline{Y} = L^2(\Omega)$

In this way one has the classical existence and uniqueness result

$$\forall\ f \in L^2(\Omega)\ :$$

(4.7) $\quad \exists!\ U \in \overline{X}\ :$

$$\Delta U = f$$

The power and simplicity - based on linearity and topological completion of uniform spaces - of the above classical existence and uniqueness result is obvious. This power is illustrated by the fact that the set $\overline{Y} = L^2(\Omega)$ in which the right hand terms f in (4.1) can now be chosen is much *larger* than the original $Y = C^\infty(\overline{\Omega})$. Furthermore, the existence and uniqueness result in (4.7) does not need the a priori knowledge of the structure of the elements $U \in \overline{X}$, that is, of the respective generalized solutions. This structure which gives the regularity properties of such solutions can be obtained by a further detailed study of the respective differential operators defining the PDEs under consideration, in this case, the Laplacean Δ. And in the above specific instance we obtain

(4.8) $\quad \overline{X} = H^2(\Omega) \cap H_0^1(\Omega)$

As seen above, typically for the functional analytic methods, the generalized solutions are obtained in topological completions of vector spaces of usual functions. And such completions, like for instance the various Sobolev spaces, are defined by certain linear partial differential operators which may happen to *depend* on the PDEs under consideration.

In the above example, for instance, the topology on the space X obviously *depends* on the specific PDE in (4.1). Thus the topological completion $\overline{X}$ in which the generalized solutions U are found according to (4.7), does again *depend* on the respective PDE.

5. Pull-Back Partial Order

Without loss of generality, [3], we shall assume that all the posets considered are without a minimum or maximum element. Various notions and results related to partial orders which are used in the sequel are presented in the Appendix.

Given an equation (3.1), (3.2), we define on X the equivalence relation $\approx_T$ by

(5.1) $\quad u \approx_T y \quad \Longleftrightarrow \quad T(u) = T(v)$

for $u, v \in X$. In this way, by considering the quotient space

(5.2) $\quad X_T = X/\approx_T$

we obtain the *injective* mapping

(5.3) $\quad T_\approx : X_T \longrightarrow Y$

defined by

(5.4) $\quad X_T \ni U \longmapsto T(u) \in Y$

where $u \in U$, that is, U is the $\approx_T$ equivalence class of u in X_T, while $T(u)$ is defined by (3.2).

At that stage, we can define a partial order $\leq_T$ on X_T as being the *pull-back* by the mapping $T_\approx$ in (5.3) of the given partial order $\leq$ on Y, namely

(5.5) $\quad U \leq_T V \quad \Longleftrightarrow \quad T_\approx(U) \leq T_\approx(V)$

for $U, V \in X_T$. The effect of the above construction is that we obtain the *order isomorphic embedding*, or in short OIE

(5.6) $\quad T_\approx : X_T \longrightarrow Y$

As mentioned, without loss of generality we shall assume that the poset $(X_T^\#, \leq_T)$ has no minimum or maximum.

And now, we consider the *order completions* $X_T^\#$ and $Y^\#$ of X_T, and respectively, Y.

For simplicity, we shall denote by $\leq$ the partial orders both on $X_T^\#$ and $Y^\#$. In fact, as seen in (A.20), these partial orders are the usual inclusion relations $\subseteq$ among subsets of X, respectively, of Y.

Then according to Proposition A.1 in the Appendix, we obtain the commutative diagram of OIE-s

(5.7)

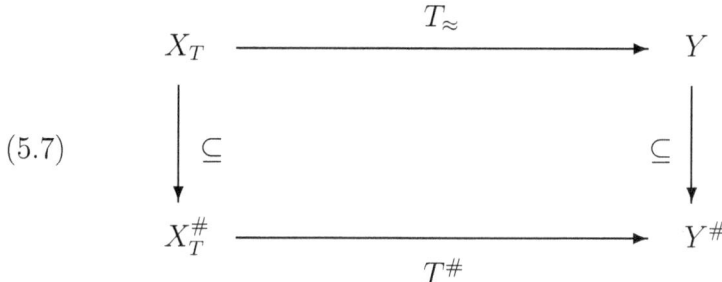

Consequently, for $U \in X_T$ and $A \in X_T^\#$, we have in $Y^\#$ the relations

(5.8) $\quad T^\#(< U]) \;=\; < T_\approx(U)]$

(5.9) $\quad T^\#(A) \;=\; (T_\approx(A))^{ul} \;=\; \sup_{Y^\#} \{\ < T_\approx(U)] \ |\ U \in A\ \}$

6. Reformulation

Now we can reformulate the problem of solving the general equations (3.1), (3.2) as follows. Given $F \in Y^\#$, find *necessary and sufficient* conditions for the existence of $A \in X_T^\#$, such that

(6.1) $\quad T^\#(A) \;=\; F$

7. Solution

We note that (5.7) gives the inclusions

(7.1)
$$\sup_{Y^{\#}} \{ T^{\#}(U) \mid U \in X_T^{\#}, T^{\#}(U) \subseteq F \} \subseteq$$
$$\subseteq T^{\#}(\sup_{X_T^{\#}} \{ U \mid U \in X_T^{\#}, T^{\#}(U) \subseteq F \}) \subseteq$$
$$\subseteq T^{\#}(\inf_{X_T^{\#}} \{ V \mid V \in X_T^{\#}, F \subseteq T^{\#}(V) \}) \subseteq$$
$$\subseteq \inf_{Y^{\#}} \{ T^{\#}(V) \mid V \in X_T^{\#}, F \subseteq T^{\#}(V) \}$$

Indeed, the first and last inclusions follow from Lemma A.1 in the Appendix. As for the middle inclusion in (7.1), let $U, V \in X_T^{\#}$ be such that $T^{\#}(U) \subseteq F \subseteq T^{\#}(V)$. Then $T^{\#}(U) \subseteq T^{\#}(V)$, hence $U \leq V$, since $T^{\#}$ is an OIE. It follows that

$$\sup_{X_T^{\#}} \{ U \mid U \in X_T^{\#}, T^{\#}(U) \subseteq F \} \leq$$
$$\leq \inf_{X_T^{\#}} \{ V \mid V \in X_T^{\#}, F \subseteq T^{\#}(V) \}$$

and the proof of (7.1) is completed.

We note further that the above inequality $T^{\#}(U) \subseteq F \subseteq T^{\#}(V)$ also implies

(7.2)
$$\sup_{Y^{\#}} \{ T^{\#}(U) \mid U \in X_T^{\#}, T^{\#}(U) \subseteq F \} \subseteq F \subseteq$$
$$\subseteq \inf_{Y^{\#}} \{ T^{\#}(V) \mid V \in X_T^{\#}, F \subseteq T^{\#}(V) \}$$

Furthermore

(7.3) $\{ U \in X_T^{\#} \mid T^{\#}(U) \subseteq F \} \neq \phi$

since (A.9) gives $U = \phi \in X_T^{\#}$, hence in view of (A.4) - (A.6) and (5.9) we have $T^{\#}(U) = (T(\phi))^{ul} = \phi^{ul} = (\phi^u)^l = (Y^{\#})^l = \phi \subseteq F$.

Returning now to the problem (6.1), we note that it is not trivial. Indeed, the OIE in (5.7), namely

$$T^{\#} : X_T^{\#} \longrightarrow Y^{\#}$$

need *not* be surjective. Further $T^{\#}$ need *not* preserve infima or suprema.

However, in the next theorem we can obtain the following two general results :

- a necessary and sufficient condition for the solvability of (6.1), and
- the explicit expression of the solution, when it exists.

Theorem 7.1.

Given $F \in Y^{\#}$.

1) The equation

(7.4) $\quad T^{\#}(A) = F$

has a solution $A \in X_T^{\#}$, if and only if, see (7.1)

(7.5)
$$\sup\nolimits_{Y^{\#}} \{\, T^{\#}(U) \mid U \in X_T^{\#},\ T^{\#}(U) \subseteq F \,\} =$$
$$= \inf\nolimits_{Y^{\#}} \{\, T^{\#}(V) \mid V \in X_T^{\#},\ F \subseteq T^{\#}(V) \,\}$$

2) This solution is unique, whenever it exists, see (5.7).

3) When it exists, the unique solution $A \in X_T^{\#}$ is given by

(7.6)
$$A = \sup\nolimits_{X_T^{\#}} \{\, U \in X_T^{\#} \mid T^{\#}(U) \subseteq F \,\} =$$
$$= \inf\nolimits_{X_T^{\#}} \{\, V \in X_T^{\#} \mid F \subseteq T^{\#}(V) \,\}$$

and, see (7.3)

(7.7) $\quad \{\, U \in X_T^{\#} \mid T^{\#}(U) \subseteq F \,\},\ \{\, V \in X_T^{\#} \mid F \subseteq T^{\#}(V) \,\} \neq \phi$

Proof

From (7.4) follows that $A \in X_T^{\#}$, $T^{\#}(A) \subseteq F$, thus

$$F = T^\#(A) \subseteq \sup_{Y^\#} \{ T^\#(U) \mid U \in X_T^\#, T^\#(U) \subseteq F \}$$

Similarly we have

$$\inf_{Y^\#} \{ T^\#(V) \mid V \in X_T^\#, F \subseteq T^\#(V) \} \subseteq T^\#(A) = F$$

thus (7.1) collapses to the seven equalities

$$F = T^\#(A) = \sup_{Y^\#} \ldots = T^\#(\sup_{X_T^\#} \ldots) =$$
$$= T^\#(\inf_{X_T^\#} \ldots) = \inf_{Y^\#} \ldots = T^\#(A) = F$$

Thus in particular we obtain (7.5).

The injectivity of $T^\#$ will give (7.6), while (7.7) follows from (7.3) and the fact that we can take $V = A$.

Conversely, let us assume (7.5). Then (7.1) collapses to the three equalities

$$\sup_{Y^\#} \ldots = T^\#(\sup_{X_T^\#} \ldots) = T^\#(\inf_{X_T^\#} \ldots) = \inf_{Y^\#} \ldots$$

thus in view of the corresponding collapsed version of (7.2), we can extend the above three equalities to the following four

$$\sup_{Y^\#} \ldots = T^\#(\sup_{X_T^\#} \ldots) = T^\#(\inf_{X_T^\#} \ldots) = \inf_{Y^\#} \ldots = F$$

And now the injectivity of $T^\#$ will give (7.4) and (7.6), while (7.7) follows as above.

□

The above existence result is of a "local" nature, since it refers to a solution of the equation (6.1) for one given right hand term $F \in Y^\#$. This result, however, can further be strengthened by the following "global" one which characterizes the solvability of (6.1) for *all* right hand terms $F \in Y^\#$. Namely, we have, [3, pp. 190,191]

Theorem 7.2.

The following are equivalent

(7.8) $\quad T^\#(X_T^\#) \supseteq Y$

and

(7.9) $\quad T^\#(X_T^\#) = Y^\#$

In each of these cases $T^\#$ is an order isomorphism, or in short, OI, between $X_T^\#$ and $Y^\#$.

It is *important* to note the following four facts :

- "Pull-back" type structures are customary when solving PDEs by functional analytic methods. Details in this regard are presented in [3, chap. 12], while one well known classical example has been presented in section 4 above.

- As shown in [3, chap. 13], and indicated in section 8 next, one can consider in (5.7) far more general partial orders than the "pull-back" type ones, and thus still obtain solutions by the order completion method for nonlinear systems of PDEs of type (1.1), together with possibly associated initial and/or boundary value problems.

- Nonlinear systems of PDEs with equations of type (1.1), together with possibly associated initial and/or boundary value problems can, in view of section 2 above, be dealt with as equations (3.1), (3.2), and therefore (6.1). Consequently, the *existence* of solutions of such systems of PDEs follows from Theorems 7.1 and 7.2.

- The *blanket, type independent*, or *universal regularity* property of such solutions, namely, that they can be associated with usual *measurable*, or even *Hausdorff continuous* functions follows from arguments in [3, 1, 4-6].

8. Beyond "Pull-Back" Partial Orders

As indicated here, and presented in full detail in [3, chap. 13], the order completion method in solving large classes of nonlinear systems of PDEs of the type (1.1) is *not* limited to the use of "pull-back" type partial orders in (45.7), (2.3) and (2.8). In fact, a large class of more general partial orders can be defined on the domains $\mathcal{M}_T^m(\Omega)$ of the respective PDEs, and stil obtain for them solutions in the corresponding order completions.

As a brief indication in this regard, let us extend the approach in section 5 above as follows. Given any equation (3.1), (3.2)

(8.1) $\quad T(A) = F$

where

(8.2) $\quad T : X \longrightarrow Y$

is any mapping, X is any nonvoid set, while $(Y, \leq)$ is a poset, while $F \in Y$ is given, and $A \in X$ is the sought after solution.

Further, let Z be any nonvoid set together with a *surjective* mapping, see (5.2)

(8.3) $\quad \lambda : Z \longrightarrow X_T$

Then we obtain the commutative diagram

(8.4)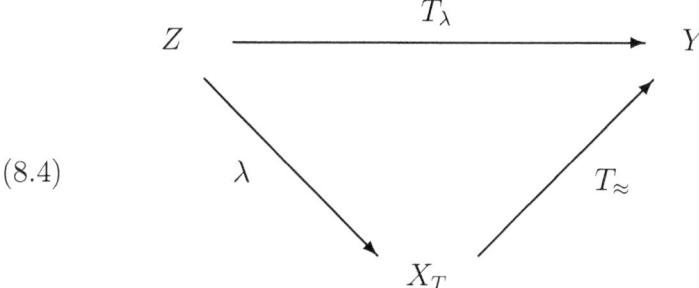

where $T_\approx$ is given in (5.3), (5.4), while we have taken by definition

(8.5) $\quad T_\lambda \;=\; T_\approx \circ \lambda$

And now, we can define define the partial order $\leq_{T,\,\lambda}$ on Z by

(8.6) $\quad z \leq_{T,\,\lambda} z' \quad \Longleftrightarrow \quad \left| \begin{array}{l} z = z', \text{ or} \\ T_\lambda(z) \lneq T_\lambda(z') \end{array} \right.$

In this case, we obtain

(8.6) $T_\approx$ is an OIE, T_λ is increasing, λ is increasing and surjective

and the procedures in sections 5 - 7 above can be applied to the mapping

(8.7) $\quad T_\lambda : Z \;\longrightarrow\; Y$

instead of the initial mapping (3.2).

Appendix

We shortly present several notions and results used above. A related full presentation can be found in [3, Appendix, pp. 391-420].

Let $(X, \leq)$ be a nonvoid poset without minimum or maximum. For $a \in X$ we denote

(A.1) $\quad <a] = \{x \in X \mid x \leq a\}, \quad [a>= \{x \in X \mid x \geq a\}$

We define the mappings

(A.2) $\quad X \supseteq A \longmapsto A^u = \bigcap_{a \in A} [a> \; \subseteq X$

(A.3) $\quad X \supseteq A \longmapsto A^l = \bigcap_{a \in A} <a] \; \subseteq X$

then for $A \subseteq X$ we have

(A.4) $\quad A^u = X \iff A^l = X \iff A = \phi$

(A.5) $\quad A^u = \phi \iff A$ unbounded from above

(A.6) $\quad A^l = \phi \iff A$ unbounded from below

Definition A.1.

We call $A \subseteq X$ a *cut*, if and only if

(A.7) $\quad A^{ul} = A$

and denote

(A.8) $\quad X^{\#} = \{A \subseteq X \mid A \text{ is a cut}\} \subseteq \mathcal{P}(X)$

$\square$

Clearly, (A.4) - (A.6) imply

(A.9) $\quad \phi, X \in X^{\#}$

therefore

(A.10) $\quad X^{\#} \neq \phi$

Given $A, B \subseteq X$, we have

(A.11) $\quad A \subseteq B \Longrightarrow A^u \supseteq B^u, \quad A^l \supseteq B^l$

(A.12) $\quad A \subseteq A^{ul}, \quad A \subseteq A^{lu}$

(A.13) $\quad A^{ulu} = A^u, \quad A^{lul} = A^l$

Consequently

$$\forall\ A \subseteq X\ :$$

$$*)\ A^{ul} \in X^{\#}$$

(A.14) $\quad **)\ \forall\ B \in X^{\#}\ :$

$$A \subseteq B \Longrightarrow A^{ul} \subseteq B$$

$$B \subseteq A \Longrightarrow B \subseteq A^{ul}$$

therefore

(A.15) $\quad X^{\#} = \{A^{ul}\ |\ A \subseteq X\}$

Given $x \in X$, we have

(A.16) $\quad \{x\}^u = [x >,\quad \{x\}^l =< x],\quad [x >^l =< x],\quad < x]^u = [x >$

(A.17) $\quad \{x\}^{ul} =< x],\quad \{x\}^{lu} = [x >$

We denote for short

$$\{x\}^u = x^u,\quad \{x\}^l = x^l,\quad \{x\}^{ul} = x^{ul},\quad \{x\}^{lu} = x^{lu},\ \ldots$$

Given $A \in X^{\#}$, we have

(A.18) $\quad \phi \neq A \neq X \Longleftrightarrow \left(\begin{array}{c} \exists\ a, b \in X\ : \\ < a] \subseteq A \subseteq < b] \end{array} \right)$

We shall use the *embedding*

187

(A.19) $\quad X \ni x \stackrel{\varphi}{\longmapsto} x^{ul} = x^l = <x] \in X^\#$

We define on $X^\#$ the partial order

(A.20) $\quad A \leq B \iff A \subseteq B$

Definition 2.1.

Given two posets $(X, \leq)$, $(Y, \leq)$ and a mapping $\varphi : X \longrightarrow Y$. We call φ an *order isomorphic embedding*, or in short, OIE, if and only if it is injective, and furthermore, for $a, b \in X$ we have

$$a \leq b \iff \varphi(a) \leq \varphi(b)$$

An OIE φ is an *order isomorphism*, or in short, OI, if and only if it is bijective.

$\square$

The main result concerning order completion is given in, [2] :

Theorem (H M MacNeille, 1937)

1) The poset $(X^\#, \leq)$ is order complete.

2) The embedding $X \stackrel{\varphi}{\longrightarrow} X^\#$ in (A.19) preserves infima and suprema, and it is an order isomorphic embedding, or OIE.

3) For $A \in X^\#$, we have the order density property of X in $X^\#$, namely

(A.21)
$$A = \sup_{X^\#} \{x^l \mid x \in X, \ x^l \subseteq A\} =$$
$$= \inf_{X^\#} \{x^l \mid x \in X, \ A \subseteq x^l\}$$

$\square$

For $A \subseteq X$, we have

(A.22) $\quad A^{ul} = \sup_{X^\#} \{x^l \mid x \in A\}$

Given $A_i \in X^\#$, with $i \in I$, we have with the partial order in $X^\#$ the relations

(A.23) $\quad \sup_{i \in I} A_i = \inf \{A \in X^\# \mid \bigcup_{i \in I} A_i \subseteq A\} = (\bigcup_{i \in I} A_i)^{ul}$

(A.24) $\quad \inf_{i \in I} A_i = \sup \{A \in X^\# \mid A \subseteq \bigcap_{i \in I} A_i\} = (\bigcap_{i \in I} A_i)^{ul} =$
$= \bigcap_{i \in I} A_i$

Extending mappings to order completions

Let $(X, \leq)$, $(Y, \leq)$ be two posets without minimum or maximum, and let

(A.25) $\quad \varphi : X \longrightarrow Y$

be any mapping. Our interest is to obtain an extension

$\varphi^\# : X^\# \longrightarrow Y^\#$

For that, we first extend φ to a *larger* domain, as follows

(A.26) $\quad \varphi^\# : \mathcal{P}(X) \longrightarrow Y^\#$

where for $A \subseteq X$ we define

(A.27) $\quad \varphi^\#(A) = (\varphi(A))^{ul} = \sup_{Y^\#} \{< \varphi(x)] \mid x \in A\}$

and for any mapping in (A.25), we obtain the commutative diagram

(A.28)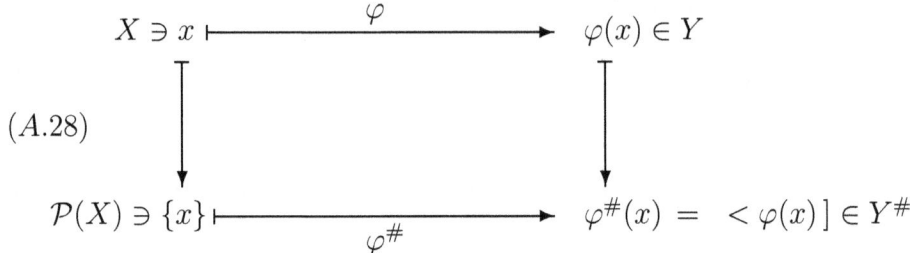

Proposition A.1.

1) The mapping $\varphi^\# : \mathcal{P}(X) \longrightarrow Y^\#$ in (A.36) is increasing, if on $\mathcal{P}(X)$ we take the partial order defined by the usual inclusion "$\subseteq$".

2) If the mapping $\varphi : X \longrightarrow Y$ in (A.35) is increasing, then the mapping $\varphi^\# : \mathcal{P}(X) \longrightarrow Y^\#$ in (A.36) is an extension of it to $X^\#$, namely, we have the commutative diagram

(A.29)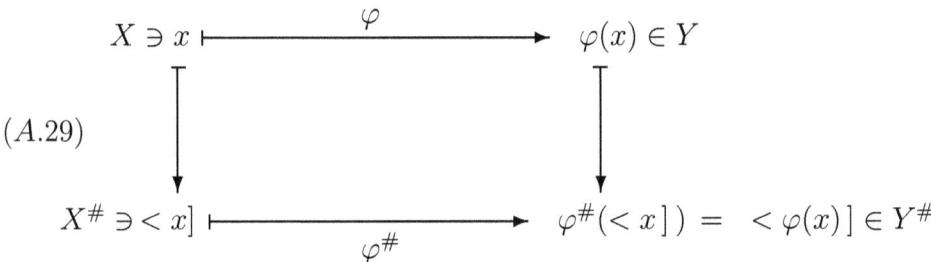

3) If the mapping $\varphi : X \longrightarrow Y$ in (A.25) is an OIE, then the mapping $\varphi^\# : \mathcal{P}(X) \longrightarrow Y^\#$ in (A.26) when restricted to $X^\#$, that is

(A.30) $\quad \varphi^\# : X^\# \longrightarrow Y^\#$

as in (A.29), is also an OIE.

Lemma A.1.

Let in general $\mu : M \longrightarrow N$ be an increasing mapping between two order complete posets, then for nonvoid $E \subseteq M$ we have

(A.31) $\quad \mu(\inf_M E) \leq \inf_N \mu(E) \leq \sup_N \mu(E) \leq \mu(\sup_M E)$

Proof

Indeed, let $a = \inf_M E \in M$. Then $a \leq b$, with $b \in E$. Hence $\mu(a) \leq \mu(b)$, with $b \in E$. Thus $\mu(a) \leq \inf_N \mu(E)$, and the first inequality is proved.

The last inequality is obtained in a similar manner, while the middle inequality is trivial.

$\square$

References

[1] Anguelov R, Rosinger E E : Hausdorff continuous solutions of nonlinear PDEs through the order completion method. Quaestiones Mathematicae, Vol. 28, 2005, 1-15, arXiv : math.AP/0406517

[2] Luxemburg W A J, Zaanen A C : Riesz Spaces, I. North-Holland, Amsterdam, 1971

[3] Oberguggenberger M B, Rosinger E E : Solution of Continuous Nonlinear PDEs through Order Completion. Mathematics Studies VOl. 181, North-Holland, Amsterdam, 1994

[4] Rosinger E E : Hausdorff continuous solutions of arbitrary continuous nonlinear PDEs through the order completion method. arXiv:math.AP/0405546

[5] Rosinger E E : Can there be a general nonlinear PDE theory for the existence of solutions ? arXiv:math.AP/0407026

[6] Rosinger E E : Solving large classes of nonlinear systems of PDEs. arXiv:math.AP/0505674

Part III : A Few Practical Suggestions ...

What to Read First, and How to Try to Read It ...

Here we shall consider *two* ways of becoming familiar, or even knowledgeable, with the order completion method in [21], when applied to solving very large classes of nonlinear systems of PDEs, with possibly associated initial and/or boundary value problems. Namely :

> 1) a more professional, thus necessarily patient and time consuming way, aimed to become knowledgeable in the subject,

and alternatively,

> 2) a possibly quick way which may hopefully still manage to lead to some familiarity with the subject.

And we shall indicate how to pursue these two ways :

a) *either* based on the book [21],

b) *or* on the present book and the literature freely available on websites.

Needless to say, getting hold of the book [21] may not be so easy. This is precisely why the present book was written.
However, for the time being, and no matter how regrettably, for a truly better understanding of the subject of the order completion method in [21], when applied to solving very large classes of nonlinear systems of PDEs, with possibly associated initial and/or boundary value problems, the book [21] is simply indispensable ...

The mentioned situation can - to some extent - be helped as follows. Pages 106-147 contain a certain amount of information on the Appendix in the book [21].

Let us now start with the way suggested for reading book [21].

And then we start with the way in 1) above.

Even here, there may - in the first iteration of reading - be an interest in a kind of "minimal time and effort" approach. And that may go as follows :

a.1) The book [21] was from the start written in such a way, as to be "user friendly". And fortunately, being thoroughly and nearly exclusively based on the use of *partial orders*, it is not supposed to involve mathematical difficulties, but rather the need to face a lack of familiarity with a variety of manifestly simple concepts, structures and properties of partial orders.

In this regard, the Appendix in [21, pp. 391-420] is absolutely essential from the very start, and as such, it cannot so far be found published anywhere else, being covered by copyright which - as usual in "good old times", and until the recent *publishing revolution* introduced by AMAZON - belongs entirely to the publisher.

The "good news" is that the mentioned Appendix can be read *in parallel* with the main text of [21], or with all the other related and mentioned papers, regarding the solutions of PDEs through the order completion method.

Regarding the book [21] itself, the following may be a useful order of reading it :

- pages VII - X

- pages 2-10

- pages 11-23 :
  focusing on (2.10), (2.12), Lemma 2.2, Proposition 2.2, Remark 2.3

- pages 24-30 :
  focusing on (3.1), (3.5), (3.8), Corollary 3.1, Remarks 3.1 and 3.2

- pages 31-37 :
  focusing on (4.1), (4.4), (4.6), Lemma 4.1, Proposition 4.1, Remark 4.1

- pages 38-64 :

focusing on Theorem 5.1

> This Theorem 5.1 is the **first** rather typical *existence* and *uniqueness* result for nonlinear PDEs of the very general form (2.1), (2.2) on page 11.
> The solution is given by (5.7), (5.10).
> The meaning of the *uniqueness* of the solutions obtained is explained on page 69.
> Theorems 5.2 and 5.3 give more powerful results.

- pages 65-73 :

> These pages are worth reading carefully, as they contain several simple, yet nontrivial examples of solving PDEs by the order completion method, as well as a few important comments of a more general interest regarding this method.

- pages 74-93 :
  focusing on

> This chapter 7 in [21] is where the **blanket, minimal, universal regularity** of solutions is presented, namely that they can *always* be assimilated with usual Lebesgue *measurable* functions on the Euclidean domains $\Omega \subseteq \mathbb{R}^n$ on which the PDEs under consideration are defined.
> As it happens, however, it is not so easy to point out what is more important at a first reading from the pages 74-93. Furthermore, being in fact a surprising and wholly *unprecedented* result regarding the *regularity* of solutions of such very large classes of nonlinear systems of PDEs, it may be worth reading the pages 74-93, seeing them as, in fact, not such a heavy price to pay ...

- pages 94-146 :
  focusing on

> The **sheaf structure** of the spaces $\hat{\mathcal{M}}_T(\Omega)^*$ which contain the *solutions* of those very large classes of nonlinear

systems of PDEs under consideration.

Here, it is important to note that, in view of all the known physical and other applicative interpretations of various PDEs encountered since the introduction by Newton of Calculus, the property to be a *sheaf* of functions of the countless spaces to which the solutions of such equations belong is essential, even if not often noted explicitly. Indeed, this sheaf property is precisely the one which allows a free and complete transition between *local* and *global* properties of such solutions, across all the open subsets $\Delta \subseteq \Omega$ of the domains of definition of the PDEs dealt with.

As it happens, the proof of the fact that the mentioned solution spaces $\hat{\mathcal{M}}_T(\Omega)^*$ do indeed turn out to be *sheaves* is not immediate, and in fact, so far, one could not find a sufficiently simple proof.

This fact seems to be inevitable, given that the spaces $\hat{\mathcal{M}}_T(\Omega)^*$ have - to our knowledge - never before been encountered in mathematics ...

On the other hand, given the special interest in the sheaf structure of the spaces $\hat{\mathcal{M}}_T(\Omega)^*$, it was found useful in [21] to present no less than *two* different proofs for that sheaf structure, one of them making use of the Axiom of Choice, while both of them turning out to be being rather long ...

- pages 146-158 :

> Present further examples, as well as counterexamples, related to spaces of solutions of the mentioned very large classes of nonlinear systems of PDEs under consideration.

- pages 161-183 :
  focusing on

> The further **surprising** property of the order completion method introduced in [21], when applied to solving very large classes of nonlinear systems of PDEs, with possibly associated initial and/or boundary value problems, namely

that the treatment of initial and/or boundary value problems does *not* introduce those considerable further technical complications and difficulties which are so typical for the functional analytic methods.

Consequently, for a first familiarization with that fact, it may be useful to read the whole of the pages 161-183.

And to give an idea of the *power* of the order completion method, on pages 182, 183, it is shown how the earlier mentioned and long outstanding Lewy *impossibility*, [12], is fully solved for the *first time* in the literature, and it is solved within a very large nonlinear margin, that is, in a much more general nonlinear version, than initially formulated. Furthermore, the solutions found are far more *regular* than mere distributions, since in view of the mentioned pages 74-93, they can be assimilated with usual measurable functions.

As a comment, let us recall the following :

The usual *functional analytic* methods for dealing with initial and/or boundary value problems for PDEs are so considerably difficult and complicated for a very simple reason, namely :

> The respective solutions are sought after in various spaces of Schwartz distributions, among them Sobolev spaces, for instance. Thus the respective distribution solutions are supposed to be *restricted* to lower dimensional *sub-manifolds* of the domains $\Omega \subseteq \mathbb{R}^n$ on which the PDEs are defined.
>
> However, as is known, restricting distributions to sub-manifolds is a highly *irregular* operation, that is, one which - by the very nature of Schwartz distributions - inevitably involves considerable technical complications, and thus altogether, difficulties ...

On the other hand - and as seen on pages 161-183 - the order completion method introduced in [21], when applied to solving very large classes of nonlinear systems of PDEs, with possibly associated initial and/or boundary value problems, does *not* encounter such considerable technical complications, and thus difficulties, when it comes to restrict elements in solution spaces to sub-manifolds.

- pages 184-194 :
  focusing on

  > A yet another **surprising** result in the order completion method, this time on the solutions of completely *arbitrary* equations :
  >
  > $T(x) = c$
  >
  > with corresponding mappings $T : X \longrightarrow Y$ restricted only by the condition that the range set Y of T be *partially ordered* in any given way.
  >
  > And for such general equations - which trivially contain the very large classes of nonlinear systems of PDEs, with possibly associated initial and/or boundary value problems considered here - one finds nothing less than *necessary and sufficient* conditions of solvability, as presented in Theorem 9.1. Furthermore, the explicit expression - in terms of partial order - of the unique solution is given in (9.18).

That may be about a first time and more professional reading of book [21] ...

And now, to the second way of reading book [21], way mentioned at 2) above, thus to :

a.2) This is rather simple to suggest, namely, as the following subset of a.1) above :

- pages VII - X

- pages 2-10

- pages 11-23 :
  focusing on (2.10), (2.12), Lemma 2.2, Proposition 2.2, Remark 2.3

- pages 24-30 :
  focusing on (3.1), (3.5), (3.8), Corollary 3.1, Remarks 3.1 and 3.2

- pages 31-37 :
  focusing on (4.1), (4.4), (4.6), Lemma 4.1, Proposition 4.1, Remark 4.1

- pages 38-64 :
  focusing on Theorem 5.1

  This Theorem 5.1 is the **first** rather typical *existence* and *uniqueness* result for nonlinear PDEs of the very general form (2.1), (2.2) on page 11.
  The solution is given by (5.7), (5.10).
  The meaning of the *uniqueness* of the solutions obtained is explained on page 69.
  Theorems 5.2 and 5.3 give more powerful results.

- pages 65-73 :

  These pages are worth reading carefully, as they contain several simple, yet nontrivial examples of solving PDEs by the order completion method, as well as a few important comments of a more general interest regarding this method.

- pages 161-183 :
  focusing on

  The further **surprising** property of the order completion method introduced in [21], when applied to solving very large classes of nonlinear systems of PDEs, with possibly associated initial and/or boundary value problems, namely that the treatment of initial and/or boundary value problems does *not* introduce those considerable further technical complications and difficulties which are so typical for the functional analytic methods.

Consequently, for a first familiarization with that fact, it may be useful to read the whole of the pages 161-183.

And to give an idea of the *power* of the order completion method, on pages 182, 183, it is shown how the earlier mentioned and long outstanding Lewy *impossibility*, [12], is fully solved for the *first time* in the literature, and it is solved within a very large nonlinear margin, that is, in a much more general nonlinear version, than initially formulated. Furthermore, the solutions found are far more *regular* than mere distributions, since in view of the mentioned pages 74-93, they can be assimilated with usual measurable functions.

- pages 184-194 :
  focusing on

> A yet another **surprising** result in the order completion method, this time on the solutions of completely *arbitrary* equations :
>
> $T(x) = c$
>
> with corresponding mappings $T : X \longrightarrow Y$ restricted only by the condition that the range set Y of T be *partially ordered* in any given way.
>
> And for such general equations - which trivially contain the very large classes of nonlinear systems of PDEs, with possibly associated initial and/or boundary value problems considered here - one finds nothing less than *necessary and sufficient* conditions of solvability, as presented in Theorem 9.1. Furthermore, the explicit expression - in terms of partial order - of the unique solution is given in (9.18).

And now, in the likely case that the book [21] is not available, we turn to a suggested way of reading of the present book, namely, to alternative b) above.

And again, we start with the way in 1) above, thus here is :

b.1) Here the following may be good to read at a first and more professional approach, and it starts with what is an adaptation of of the 2004 paper :

Rosinger E E : Can there be a general nonlinear PDE theory for the existence of solutions ? arXiv:math/0407026

Namely :

- pages 6-27, 32-53, 53-64, 73-87, 148-168

- pages 87-105 :   focusing on

> The surprising and totally **unprecedented** *blanket, universal, minimal regularity* property of solutions of the very large classes of nonlinear systems of PDEs, with possibly associated initial and/or boundary value problems considered here, according to which such solutions can always be assimilated with *Hausdorff-continuous* functions.

b.2) As for a quicker reading of the present book, we can suggest :

- pages 6-27, 32-53, 73-87, 148-168

www.ingramcontent.com/pod-product-compliance
Lightning Source LLC
Chambersburg PA
CBHW051648170526
45167CB00001B/376